Physical Properties of Materials for Engineers

Volume III

Author

Daniel D. Pollock
Professor of Engineering
State University of New York at Buffalo
Buffalo, New York

CRC Press, Inc.
Boca Raton, Florida

Library of Congress Cataloging in Publication Data

Pollock, Daniel D.
Physical properties of materials for engineers.

Includes bibliographies and indexes.
1. Solids. 2. Materials. I. Title.
QC176.P64 620.1'12 81-839
ISBN 0-8493-6200-8 (set) AACR2
ISBN 0-8493-6201-6 (v. 1)
ISBN 0-8493-6202-4 (v. 2)
ISBN 0-8493-6203-2 (v. 3)

This book represents information obtained from authentic and highly regarded sources. Reprinted material is quoted with permission, and sources are indicated. A wide variety of references are listed. Every reasonable effort has been made to give reliable data and information, but the author and the publisher cannot assume responsibility for the validity of all materials or for the consequences of their use.

Direct all inquiries to CRC Press, Inc., 2000 N.W. 24th Street, Boca Raton, Florida 33431.

International Standard Book Number 0-8493-6200-8 (Complete Set)
International Standard Book Number 0-8493-6201-6 (Volume I)
International Standard Book Number 0-8493-6202-4 (Volume II)
International Standard Book Number 0-8493-6203-2 (Volume III)

Library of Congress Card Number 81-839
Printed in the United States

PREFACE

Many new materials and devices which were designed to possess specific properties for special purposes have become available in the recent past. These have had their origins in basic scientific concepts. Engineers must understand the bases for these developments so that they can make optimum use of available materials and further advance the existing technology as new materials appear. The main objective of this text is to provide engineers and engineering students a unified, elementary treatment of the basic physical relationships governing those properties of materials of greatest interest and utility.

Many texts on solid state physics, written primarily for advanced undergraduate physics courses, make use of sophisticated mathematical derivations in which only the most significant parts are given; the intermediate steps are left to the reader to provide. This makes it difficult for the average engineer to follow and has the effect of discouraging or "turning off" many readers. Other texts are not much more than surveys of "materials science" and provide little insight into the nature of the phenomena.

This text represents an attempt to provide a middle ground between these extremes. It is designed to explain the origin and nature of the most widely used physical properties of materials to engineers; thus, it prepares them to understand and to utilize materials more effectively. It also may be used as a textbook for senior undergraduate and first-year graduate students.

Practicing engineers will find this text helpful in getting up to date. Readers with some familiarity with this field will be able to follow the presentations with ease. Engineering students and those taking physics courses will find this book to be a useful source of examples of applications of the theory to commercially available materials as well as for uncomplicated explanations of physical properties. In many cases alternate explanations have been provided for clarity.

An effort has been made to keep the mathematics as unsophisticated as possible without "watering down" or distorting the concepts. In practically all cases only a mastery of elementary calculus is required to follow the derivations. All of the "algebra" is shown and no steps in the derivations are considered to be obvious to the reader. Explanations are provided in cases where more advanced mathematics is employed The problems have been designed to promote understanding rather than mathematical agility or computational skill.

The introductory chapters are intended to span the gap between the classical mechanics, which is familiar to engineers and engineering students, and the quantum mechanics, which usually is unfamiliar. The limitations of the classical approach are shown in elementary ways and the need for the quantum mechanics is demonstrated. The quantum mechanics is developed directly from this by the use of uncomplicated examples of various phenomena. The degree to which the quantum mechanics is presented is sufficient for the understanding of the physical properties discussed in the subsequent chapters; it also provides a sound basis for more advanced study.

Introductory sections are given which guide the reader to the topic under consideration. The basic physical relationships are provided. These are drawn from concepts and properties which are known to those with engineering backgrounds; they lead the reader into the topic of interest. In some cases small amounts of material are repeated for the sake of clarity and convenience. Some topics, frequently presented as separate chapters in physics texts, have been incorporated in various sections in which they are directly applicable to materials. Lattice dynamics is one of the subjects treated in this way. Where appropriate, sections covering the properties of commercially available materials are included and discussed. This approach provides more comprehensive presentations which can be readily followed and applied by the reader.

Since this text is intended for readers with engineering backgrounds, some of the topics often presented in solid state physics books have been omitted. The reader is, however, provided with a suitable foundation upon which to pursue such topics elsewhere. The fundamentals of solid state physics are indispensable to the understanding of the properties of materials; these have been retained. Thus, the approach and content of this text are unique in that they include the properties and applications as well as the theory of those major types of real materials which are most frequently employed by engineers. This is rarely, if ever, done in current physics texts.

On the other hand, important subjects marginally included, or omitted, from many physics texts have been incorporated. Chapter 6 (Electrical Resistivities and Temperature Coefficients of Metals and Alloys) and Chapter 7 (Thermoelectric Properties of Metals and Alloys) are good examples of this. These chapters are unique in that similar material does not appear in any text of which I am aware. Sections of these chapters include the basic physical theories and their relationships to phase equilibria as well as their application to the design of alloys with special sets of electrical properties and to the explanations of the properties of commercially available alloys. The very wide use of these types of alloys makes it necessary that engineers thoroughly understand the mechanisms responsible for their optimum applications and their limitations. Other topics of primary importance to engineers which are normally included in solid state physics texts also have been incorporated.

The background required for this text includes elementary calculus, first-year, college-level physics and chemistry, and one course in physical metallurgy or materials science. Information required beyond these levels has been incorporated where needed. This makes it possible to accommodate the needs of readers where there is a wide range of background and capability; it also permits self-study.

The first five chapters introduce, explain, and develop the modern theory of solids; these are considered to constitute the minimum basis for any text of this type. Various other sections, or chapters, may then be studied, depending upon the interests of the reader and the emphasis desired. One combination of topics could be selected by electrical engineers, another set by metallurgical engineers, still another group by mechanical engineers, etc. Courses in materials engineering could be organized in similar ways. It should be noted, however, that all of the major topics included in this text represent physical properties employed by, and of significance to, most engineers at some time during their careers.

Note should be made that the units used in each of the topics are those currently employed by engineers working with materials in that area. The use of a single system of units would be counterproductive. Means for conversions to other units are given in the text for convenience and in the appendix.

I wish to express my deep appreciation to two of my former teachers for the insights and approaches to solid state phenomena which they provided early in my career. Professor C. W. Curtis, of Lehigh University, and Dr. F. E. Jaumot, then associated with the University of Pennsylvania, have been continuing sources of inspiration. In addition, some of the illustrations given in Dr. Curtis' lectures have served, with his permission, as models for the equivalents given here. Similarly, I am indebted to Dr. Jaumot for permission to use his clear approaches to reciprocal space, Brillouin zone theory, and the elementary theory of alloy phases as a basis for those used here.

I am deeply grateful to the American Society for Testing and Materials for permission to condense the contents and to use the illustrations from the monograph, *The Theory and Properties of Thermocouple Elements*, STP492, 1971, written by the author. This material is presented in Chapter 7 (Volume II).

Acknowledgment is also made of the assistance provided by Mr. James Stewart for his cooperation and assistance in the preparation of the illustrations.

I am very grateful to Donna George for her unfailing patience and help in typing the manuscript.

Credits are given with the individual tables and figures.

Daniel D. Pollock

PHYSICAL PROPERTIES OF MATERIALS FOR ENGINEERS

Daniel D. Pollock

Volume I

Volume II

Volume III

TABLE OF CONTENTS

Volume III

Chapter 10

PHYSICAL FACTORS IN PHASE FORMATION

Thermodynamic criteria, such as the Gibbs free energy of formation, give the probability of the appearance of a new phase. However, they provide incomplete, basically chemical models and no physical mechanisms to explain phaseal relationships. It is for this reason that many thermodynamic approaches are insufficient by themselves.

While relatively little has been accomplished, as yet, by means of physical approaches, it is believed that this will play a very prominent role in the future. Neither point of view is sufficient unto itself. When both the thermodynamics and physics are employed together, the combination yields greater understanding than either one by itself. Some of the physical aspects are given here. This chapter is not intended to include all types of alloy phases; some of the more common types are included to provide an insight into the physical factors involved in phase formations.

10.1. INTERSTITIAL SOLID SOLUTIONS AND COMPOUNDS

Interstitial solid solutions are those in which the alloying ions take positions in the spaces, or interstices, between the host ions which are on regular lattice sites. It is natural to assume that the size of an ion which would fit into an interstitial position is limited by the sizes of the voids between the ions occupying normal lattice sites. Thus, it is expected that the geometry of the lattice would be an important factor in the formation of these solutions.

Another factor to be considered is that it is probable that ions occupying interstitial lattice sites rarely would be of sizes which would fit exactly into the lattice voids without distorting the lattice. It is, therefore, expected that interstitial ions would have a high probability of distorting the host lattice and increasing its energy. Thus, only those interstitial sites would tend to be occupied which would cause the least amount of distortion of the surrounding portions of the host lattice. This constitutes another limitation upon the capacity of a lattice to retain ions in interstitial sites. In addition, the ability of the host lattice to accommodate the accompanying strains is limited. This imposes another constraint upon the ability of a lattice to maintain interstitial ions in solution.

In addition to the disruption of the spatial periodicity of the host lattice, its electrical periodicity also is disturbed (Equation 6-19a, Volume II). This can affect the bonding between the ions. Such changes may be minimized if the bonding energies of the two species of ions are similar; the greater their difference, the smaller is the probability that a solution of this kind will be formed. In this case, the probability is high that another phase, rather than an interstitial solution, will be formed.

Elements forming interstitial solid solutions can have important effects upon the properties of metals and alloys. Hydrogen embrittlement (large decreases in toughness and ductility) can result from relatively small amounts of hydrogen in interstitial sites, particularly in ferrous alloys. The hydrogen may enter an alloy as a result of cleaning processes involving acids, electroplating, or improper welding or melting techniques.

Strain aging, a phenomenon which can have undesirable effects in successive, deep-drawing operations with low-carbon steels, may be caused by carbon or by nitrogen which is introduced unintentionally during the melting process. Interstitial elements such as these also adversely affect the magnetic properties of high-permeability, ferrous, magnetic alloys. The mechanical properties of titanium and its alloys also are degraded by the presence of interstitial C, H, and N.

Another element in this class, boron, has an extremely strong effect in increasing the hardenability of steels. This is true even when it is present in very small amounts. The boron content of such steels is carefully controlled for this reason. The maximum boron content usually is maintained at levels well below 0.001 wt %.

Most important of all, the control of carbon, in and out of interstitial solid solution in iron, defines the heat treatments and resulting microstructures of steels. The mechanical properties of steels are directly related to their microstructures. The ability to manipulate the microstructures of steels, and, consequently, to control their properties, greatly increases their utility.

Examples such as these demonstrate the practical engineering significance of interstitial solid solutions.

10.1.1 Hägg's Rules for Interstitial Solid Solutions

Studies of interstitial solid solutions, by Hägg, have resulted in the formulation of empirical ''rules'' governing their behavior. The first of these is that the ratio of the ''atomic diameter'' of the ion occupying an interstitial site should be equal to or less than 0.59 that of the host ion when considered as close-packed spheres. When the interstitial ion is smaller than this, FCC or HCP lattices usually are formed by the host (metal) ions. Interstitial ions exceeding this limit result in the formation of more complex lattices (see Table 10-5). On the strictly geometric basis of a FCC lattice of hard spheres, an interstitial ion occupying an octahedral position (six nearest neighbors) must have a radius greater than 0.41 that of the host ion in order to be tangent to its nearest neighbors. Interstitial ions smaller than this can occupy tetrahedral sites (four nearest neighbors) if their radii are not less than 0.23 of the host ion and they touch their nearest neighbors. Of course, certain nonclose-packed lattices can, on a comparable basis, accommodate larger ions at the appropriate positions within the lattice.

This confirms the assumption that size and geometric effects should play an important part in these solutions. Hägg also showed that those elements most likely to occupy interstitial sites are H, B, C, and N. These elements have relatively small ionic diameters; the elements most likely to accommodate such atoms interstitially are the transition elements.

The composition ranges for many of such solutions are rather limited. This results primarily from two factors. The first of these is the warpage, or strain, induced into the host lattice by the formation of such solutions. The lattice strain energy rises rapidly as a function of the amount of a solute in interstitial solution. A point is soon reached at which an additional increment of the added element induces such high strain energy that the formation of a new phase occurs because it represents a lower energy situation. This is an additional factor which tends to limit the extent of solid solubility.

Another factor limiting such solid solutions is that the differences in the electronegativities (Section 10.2.2) between the four interstitial elements previously noted and those of the transition elements are relatively large. This means that compound formation is more probable than extended solid solubility.

The electronegativity differences between these classes of elements are sufficiently large to severely limit the ranges of solubility in most cases. As can be seen in Table 10-1, most of the combinations show relatively little solid solubility. The tendency is one of compound formation. It also is interesting to note that almost all of the resulting compounds are close-packed phases (see Section 10.2.1.2).

Hägg also noted that this class of compounds is essentially metallic in nature when transition elements are components of the compounds. These show metallic electrical conductivity and have a metallic luster. Nonmetallic behavior is shown when such compounds contain normal metals rather than transition elements. (It will be noted that

Table 10-1
INTERSTITIAL ALLOYS

System	Lattice type	Solute radius/ solvent radius	Maximum α solid solubility	Limiting compound
Zr-H	HCP	0.29	50	Zr_2H
Ta-H	HCP	0.32	30	Ta_2H
Ti-H	HCP	0.32	7.9	Ti_2H
Ta-C	HCP	0.53	3	Ta_2C
Mn-N	HCP	0.55	0.5	Mn_2N
W-C	HCP	0.55	Insoluble	W_2C
Cr-N	HCP	0.56	Negligible	Cr_2N
Mo-C	HCP	0.56	Insoluble	Mo_2C
Fe-N	HCP	0.56	0.1	Fe_2N
V-C	HCP	0.58	∼1	V_2C
Ta-H	BCC	0.32	30	TaH
Mo-N	Simple Hex.	0.52	?	MoN
W-C	Simple Hex.	0.58	Insoluble	WC
Zr-H	FCC	0.29	∼50	Zr_4H
Ti-H	FCC	0.32	7.9	Ti H
Pd-H	FCC	0.34	5	PdH_2
Zr-N	FCC	0.45	4.4	ZrN
Sc-N	FCC	0.47		ScN
Zr-C	FCC	0.48		ZrC
Nb-N	FCC	0.49	0.05	NbN
Ti-N	FCC	0.49	∼21	TiN
W-N	FCC	0.49	Insoluble	W_2N
V-N	FCC	0.53	0.7	VN
Nb-C	FCC	0.53	0.2	NbC
Ti-C	FCC	0.53	7	TiC
Ta-C	FCC	0.53	3.	TaC
Mn-N	FCC	0.55	0.5	Mn_4N
Cr-N	FCC	0.56	Negligible	CrN
Fe-N	FCC	0.56	0.1	Fe_4N
V-C	FCC	0.58	∼1	VC
Fe-C	BCC	0.63	0.02	Fe_3C

Based on Seitz, F., *The Physics of Metals,* McGraw-Hill, New York, 1943, 40. Also see References 3 and 4.

all of the compounds shown in Table 10-1 contain transition elements.) Those with the normal metals, such as CaC_2, form a slightly distorted NaCl lattice and some are transparent. Those with boron sometimes form complex structures.

Interstitial compounds frequently have the compositions given by MA, M_2A, and M_4A, in which M is the metal ion and A is the interstitial ion. Some of these compounds may exist over a range of compositions as a result of the extent to which the interstitial sites are filled.

Many of these compounds with transition elements are very hard. Use is commonly made of this in the nitriding or carburizing of steels to improve their wear resistance. High-speed steels, which can maintain their cutting edges at relatively high temperatures because of such compounds, are an important factor in mass production. Other carbides, such as WC, are cemented together to form high-speed cutting tools.

10.2. SUBSTITUTIONAL SOLID SOLUTIONS

The solute ions occupy regular lattice sites in this type of solid solution, i.e., they take the place of host ions which would occupy such positions normally. It would be expected that the presence of alloying ions in such a solid solution would also affect the host lattice. This is the case, but the influence of substitutional ions usually is less pronounced that that of interstitial ions. To continue the geometric reasoning, it can be seen that ranges of solid solubility of such solutions should, in all probability, be of greater extent than those of interstitial solutions because the lattice distortion should, in general, be smaller. It also is expected that the change in the lattice parameter of the host lattice should be a function of the amount of alloy present in solution; the more solute ions in solution in the host lattice, the greater the expected change in the lattice parameter.

Generally, increases in the amount of the second element change the lattice parameter in a regular way. It would be expected that ions smaller than the host would have a tendency to diminish the lattice parameter of the solution; larger ions would tend to increase this. It also would be expected that if the size difference between the ions were very small, virtually no change would occur in the lattice parameter of the alloy. This is seldom the case, but does occur. The most frequently observed change is one in which the lattice parameter as a function of the amount of alloying element shows a positive slope. Some systems show little change in lattice parameter (Figure 10-1). Such behavior also is shown by the Ni-Co system. Systems with decreasing lattice parameters as a function of alloy content also are common.

The linear behavior of the lattice parameter as a function of composition is known as Vegard's law. It originally was based upon data from solid solutions in ionic crystals. This rule should be regarded as a first approximation of the behavior of the most metallic solid solutions. Many of these do not show such simple, linear behavior over extended ranges of composition.

However, even assuming Vegard's rule as being operative, the same major factors influencing interstitial solid solutions should also affect substitutional solid solutions as well. Here, even though size differences between the solvent and solute ions usually are smaller, lattice strain effects as well as differences in bond strengths would be expected to be present. Certainly the shape of the Brillouin zone, as well as the Fermi level (Section 7.9, Volume II), would be expected to change. All of these factors would be expected to have a strong influence upon the extent of solid solubility.

Other less obvious factors influence solid solubility in addition to those cited above and are discussed subsequently. No one of these can be used as a criterion by itself. All of the factors must be considered in conjunction with one another to obtain the most complete understanding of the behavior of solid solutions.

10.2.1. Size Factors

From the foregoing, size, or geometric, effects are expected to influence the extent of solid solubility. Hume-Rothery was the first to recognize this. His empirical rule for this states that when the "sizes" of the host and solute "atoms" differ by more than 15%, solid solubility is restricted. When the ion "diameters" differ by less than 15%, solid solubility is favored and extended ranges of solubility frequently occur. This follows from the idea that the better the size match of the ions, the smaller will be the distortion and strain energy of the host lattice and the greater will be the capacity of the solvent lattice to accommodate the alloying ions.

The "atomic diameter" used by Hume-Rothery is the distance of closest approach of the ions in the pure element. This is not the lattice parameter. For example, in the BCC lattice, the distance of closest approach is equal to one-half of the cube diagonal;

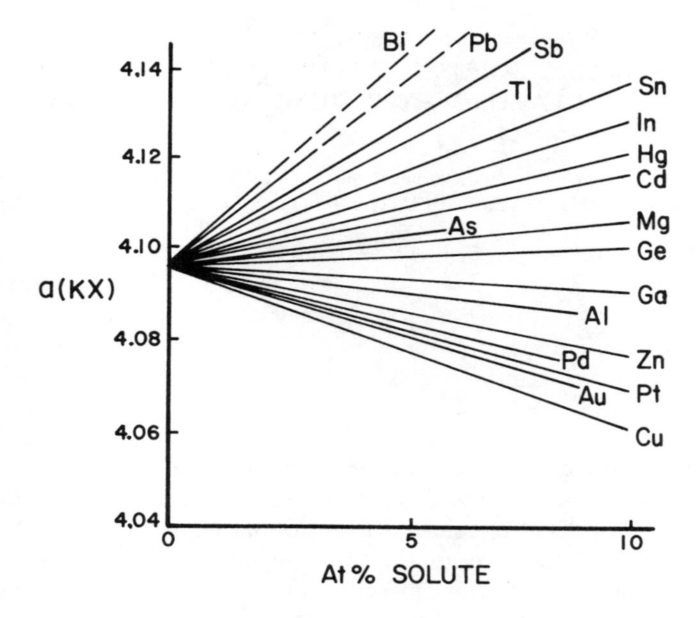

FIGURE 10-1. Lattice parameters of solid solutions in silver. (After Pearson, W. B., *Handbook of Lattice Spacings,* Pergamon Press, Elmsford, N.Y., 1958, 263. With permission.)

in the FCC lattice this is half of the face diagonal. This basis of comparison of diameters has the advantage of not requiring an *a priori* knowledge of the crystal structure of the alloy system involved; it can be misleading.

Difficulties with this size criterion can arise when the degree of ionization of the solute ion changes. Indications are that ionic size changes can vary with the quantity of solute, the kind and quantity of any other solute ions in solution, as well as the formation of super-lattices, or ordered structures, and/or compounds. It should be emphasized that "atomic size" is not necessarily a constant for a given element, and that it can show considerable variation.

Despite such variations, the Hume-Rothery rule has been shown to be very discriminating in the prediction of cases of restricted solid solubility. Calculations based upon elasticity theory show that the lattice strain energy induced by the presence of a larger ion on a lattice site is related to the 15% size difference.

Since other factors also exert strong influences upon solid solubility, this important rule should not be used as the sole criterion. This is especially true for cases of extended ranges of solid solubility.

10.2.1.1. Goldschmidt Radii

Because of the previously noted difficulties encountered in the determination of "atomic size", it frequently is convenient to use what is known as the Goldschmidt atomic radius. This is based upon the number of nearest neighbors, or the coordination number. For example, atoms crystallizing in the FCC system are close-packed and have a coordination number of 12 (CN 12), those in the BCC system (nonclose-packed) have CN 8, or eight nearest neighbors.

The Goldschmidt radius of a given element, thus, is expected to vary with any allotropic changes. The radii of such elements in closed-packed lattice structures appear to be slightly larger than when in a nonclose-packed crystal structure. Compared to a CN 12 structure, the radius of an element with CN 4 is about 12% smaller, CN 6 is about 4% smaller, and CN 8 is about 3%smaller than when in the close-packed array.

Table 10-2
PROBABLE SOLUBILITIES OF SOME ELEMENTS IN Cu AND Ag USING THE HUME-ROTHERY RULE

Alloying element CN 12 radius		% difference from radius of Cu (1.28)	% difference from radius of Ag (1.44)	Solubility	
				Cu	Ag
Zn	1.37	7	7	+	+
Mg	1.60	53	10	−	+
Cd	1.52	19	6	−	+
Hg	1.55	21	8	−	+
Be	1.13	12	22	+	−

Several elements do not always appear in close-packed arrays; many only form non-close-packed lattices. Examples of this type are the metallurgically important elements Mo and W. It is convenient to compare ion sizes on the basis of a given coordination number. This approach may be used only for metallic bonding, since in addition to the other previously noted factors affecting ionic sizes, ionic, covalent, and mixed bonding can have large effects upon the apparent size.

The Hume-Rothery size criterion, applied to CN 12 radii, is used with some pairs of elements in Table 10-2. These are provided to illustrate the Hume-Rothery method of predicting the probable extent of the terminal solid solutions formed by these elements with Cu and Ag. The plus signs indicate that a reasonable range of solubility is predicted; the minus signs indicate that limited solubility is probable. The closer the atom sizes are, the more probable is the formation of an extended terminal solid solution. The greater the disparity between sizes, the less probable is the formation of an extended α solid solution. When the size factor is close to 15%, the prediction is uncertain. It should be understood that this illustrates a general principle: if the size factor is unfavorable, the extent of the terminal solid solution probably will be small. Data for various radii are given in Table 10-3.

The size factor for ternary alloys may be calculated using the equation

$$D_{ABC} = \frac{C(B)D_{AB} + C(C)D_{AC}}{C(B) + C(C)}$$

in which

$$D_{IJ} = \frac{r_I - r_J}{r_I} \times 100$$

gives the size factor for a binary system. The radii r_A, r_B, and r_C may be taken either as one-half of the distance of closest approach, or obtained by other means. The components, C(B) and C(C), are expressed in atom percent.

10.2.1.2. Space Concept

Laves has defined another factor known as the "space principle". Neglecting those elements in the periodic table in the columns headed by C, N, O, and F, 58 of the elements form close-packed crystal structures, either FCC or HCP lattices with CN 12. The ideal HCP lattice has a c/a ratio of 1.63 for closest packing. These are the most efficient packings of hard spheres. This maximum filling of space introduces another geometrical factor. Elements crystallizing in these lattices must form terminal solutions which also conform to the optimum filling of space. The tendency to preserve optimum space filling is the "space principle"; this implies that maximum density is favored.

This behavior is not limited to solid solutions. Most of the compounds shown in Table 10-1 crystallize in either HCP or FCC lattices. Packing densities greater than CN 12 are found where ions with correct size differences are involved. For example, Cr_3Si structures may have CN 14. This adds emphasis to the space principle.

10.2.1.3. Symmetry Concept

The BCC lattice type is the next most prevalent structure. Of the remaining elements, 33 fall in this class. Elements with this structure can be classified as shown below:[5]

1. Allotropic elements that have close-packed structures at low temperature and BCC structures at high temperatures: Li, Na, Ca, Sr, Th, Zr, Ti, Tl (Fe) and (Mn).
2. Allotropic elements of reverse behavior with BCC structures at low temperatures and close-packed structures at high temperatures: Cr, Fe.
3. Uranium with complex structures at low temperature and BCC structure at high temperature (also Mn).
4. Elements for which only BCC structures have been found: K, Rb, Cs, Ba, Eu, V, Nb, Ta, Mo, W.

The FCC lattice represents the most effective filling of space by hard spheres of constant size. Elements of the first type given above decrease from CN 12 to CN 8, a less efficient packing. On the basis of geometry alone, other structures between these two are possible. A structure with CN 10, for example, could fill the space more effectively than one of CN 8. However, such lattices are not observed. The BCC lattice (CN 8), which is found, is preferred because of the high lattice symmetry. This degree of symmetry is absent in arrays with configurations intermediate between CN 8 and CN 12. The transformation from close packing to BCC induces a higher degree of directional bond distribution and symmetry than is present in the other lattice configurations. The increase in bond directionality, based on this high symmetry, and resulting in a lower energy configuration, explains the presence of the BCC array in preference to other lattice types intermediate between CN 8 and CN 12.

The second class of elements behaves in an opposite way from the first. Here, as the temperature increases, the degrees of bond directionality and symmetry diminish because of the relatively large vibrational amplitudes and the ions can fill the space more effectively at higher densities and lower energies in the form of FCC lattices, as noted in the discussion of the space principle.

The elements in the third and fourth categories must be considered to form BCC lattices primarily on the basis of bond directionality and high symmetry resulting from CN 8.

The lines of reasoning presented in this and in the previous sections have not been proven rigorously, nor are they fully acceptable to many observers. They provide bases for insight into wide ranges of experimentally observed metallurgical phenomena, however, and must be taken into account in considering solid solution and compound formation in the absence of more exact criteria. An obvious example is provided by the changes in solid solubility which can occur when allotropic transformations take place.

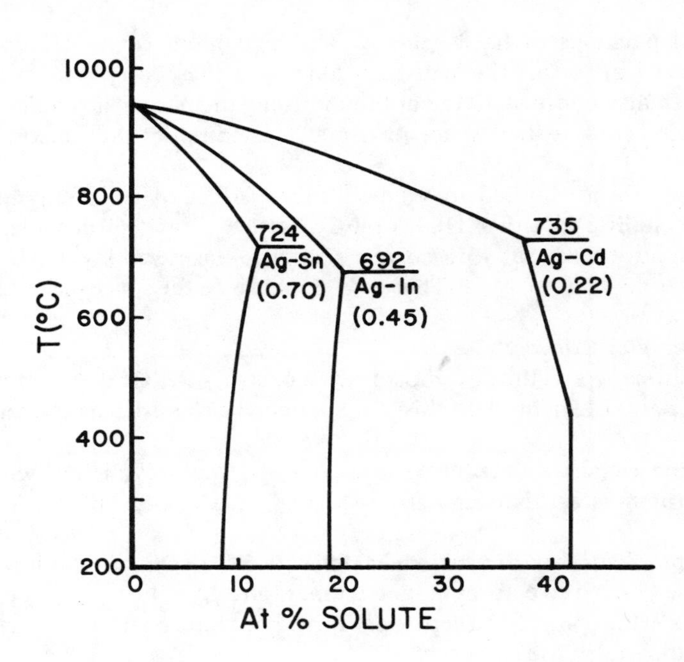

FIGURE 10-2. Extent of terminal solid solutions of some silver-base alloys (electronegativity differences from Table 10-4). (After Hume-Rothery, W., *The Structure of Metals and Alloys,* Chemical Publishing, New York, 1939, 63. With permission.)

10.2.2. Electronegativity

The more chemically active behavior that an element demonstrates, the greater would be the expectation that compounds, rather than solid solutions, would be formed by it. Pauling found that the tendency of two elements to form ionic compounds could be expressed by

$$\Delta H \text{ (kcal/g atom)} = -23.07\, Z\, (x_a - x_b)^2 \tag{10-1}$$

where ΔH is the heat of formation of the compound (the energy required to form the bonds), Z is the number of valence bonds involved in the compound, and x_a and x_b are the electronegativities of the two elements involved, in electron volts. Electronegativities also may be determined from work functions and other means. They vary slightly, depending upon the method used. These usually have an uncertainty of ±0.1 eV. Equation 10-1 can be expressed as

$$x_a - x_b = \left[\frac{\Delta H}{-23.07\, Z}\right]^{1/2} \tag{10-2}$$

Thus, the electronegativities in Equation 10-2 are in units of the square root of bond strength. The larger the difference between the electronegativities, the greater will be the probability of the formation of a stable compound.

Hume-Rothery applied this idea as an additional criterion for the determination of the probability of the formation of solid solutions. Using Pauling's findings, it was considered that the smaller the difference between the electronegativities of the elements involved, the smaller the probability of compound formation and the greater the probability of the formation of a solid solution.

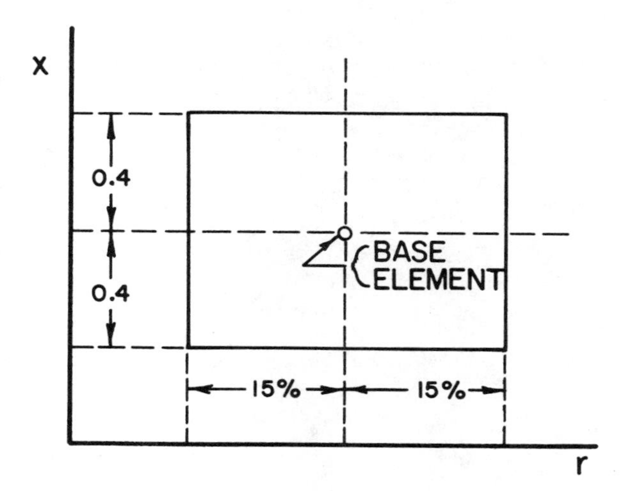

FIGURE 10-3. Plot of electronegativity and size factor.

Thus, where a favorable size factor exists, small differences in electronegativity between the two elements should favor larger ranges of solid solubility. Under similar size conditions, large electronegativity differences should favor more limited solubilities and/or compound formation.

Where a stable compound is formed at the expense of an extended terminal solution, relatively large decreases in solid solubility with decreasing temperature frequently are observed. In general, solid solubility usually is quite limited when the electronegativity difference is greater than about ±0.4 units. The effects of valence and electronegativity on the extent of the solid solubility of some silver-base alloys is given in Figure 10-2. It will be shown in later discussions that other factors also have important influences upon the extent of solid solubility.

The limiting electronegativity range given above can be used along with the Hume-Rothery size factor to obtain a better approximation than is possible by the use of either factor alone. This approach is shown in Figure 10-3. Elements which fall within the rectangular area given by both sets of limits should be good candidates for solid solutions in the base element. The closer that the points which represent solute elements are to that of the base element, the greater is the probability of extended solubility. Information on electronegativities is given in Tables 10-3 and 10-4.

It should be remembered that the Pauling idea of electronegativity is based upon ionic bonding. The metallic bond can be considered to be a very unsaturated covalent bond, but this is a great extrapolation; it can lead to contradictory results if rigorously applied to alloy phase formation. As is discussed later, other types of bonding may take place in metallic and intermetallic phases. At most, Equation 10-1 can be used to indicate a tendency toward compound formation. Used in this way, it provides a convenient parameter.

Gordy extended Pauling's ideas by considering the electronegativity in terms of the effect of the nuclear charge upon a valence electron which is in an orbit of mean distance r from it. This distance, r, is the single-bond, covalent radius. When simplifying assumptions are made regarding the screening effects of the electrons in the completed levels, it is found that the electronegativity is given by

$$x = 0.31 \left[\frac{n+1}{r}\right] + 0.50 \qquad (10\text{-}3)$$

Table 10-3
RADII AND ELECTRONEGATIVITIES OF ELEMENTS

Element	Goldschmidt radii CN 12	Goldschmidt radii CN8	Pauling electronegativity
H	(0.78)	0.76	2.1
Li	1.57	1.54	1.0
Be	1.13	1.11	1.5
B	0.95	0.93	2.0
C	0.86	0.84	2.5
N	0.8	—	3.0
Na	1.92	1.88	0.9
Mg	1.60	1.57	1.2
Al	1.43	1.40	1.5
Si	1.34	1.31	1.8
P	1.3	—	2.1
S	—	—	2.5
K	2.36	2.31	0.8
Ca	1.97	1.93	1.0
Sc	1.60	1.57	1.3
Ti	1.45	1.42	1.5
V	1.36	1.33	1.6
Cr	1.28	1.25	1.6
Mn	1.31	1.28	1.5
Fe	1.27	1.24	1.8
Co	1.26	1.23	1.8
Ni	1.24	1.22	1.8
Cu	1.28	1.25	1.9
Zn	1.37	1.34	1.6
Ga	1.39	1.36	1.6
Ge	1.39	1.36	1.7
As	1.48	1.45	2.0
Se	1.6	—	2.4
Rb	2.53	2.48	0.8
Sr	2.16	2.12	1.0
Y	1.81	1.77	1.2
Zr	1.60	1.57	1.4
Nb	1.47	1.44	1.6
Mo	1.40	1.37	1.8
Tc	1.36	1.33	1.9
Ru	1.32	1.29	2.2
Rh	1.34	1.31	2.2
Pd	1.37	1.34	2.2
Ag	1.44	1.41	1.9
Cd	1.52	1.49	1.7
In	1.57	1.54	1.7
Sn	1.58	1.55	1.8
Sb	1.61	1.58	1.9
Te	1.7	—	2.1
Cs	2.74	2.69	0.7
Ba	2.25	2.20	0.9
La	1.87	1.83	1.1

Table 10-3 (continued)
RADII AND ELECTRONEGATIVITIES OF ELEMENTS

Element	Goldschmidt radii CN 12	CN8	Pauling electronegativity
Ce	1.83	1.79	1.05
Pr	1.82	1.78	1.1
Nd	1.82	1.78	[a]
Pm	—	—	[a]
Sm	—	—	[a]
Eu	2.02	1.98	[a]
Gd	1.79	1.75	[a]
Tb	1.77	1.73	[a]
Dy	1.77	1.73	[a]
Ho	1.76	1.72	[a]
Er	1.75	1.71	[a]
Tm	1.74	1.71	[a]
Yb	1.93	1.89	[a]
Lu	1.74	1.71	[a]
Hf	1.58	1.55	1.3
Ta	1.46	1.43	1.5
W	1.41	1.38	1.7
Re	1.37	1.34	1.9
Os	1.34	1.31	2.2
Ir	1.35	1.32	2.2
Pt	1.38	1.35	2.2
Au	1.44	1.41	2.4
Hg	1.55	1.52	1.9
Tl	1.71	1.68	1.8
Pb	1.75	1.71	1.8
Bi	1.82	1.78	1.9
Po	1.8	—	2.0
At	—	—	2.2
Ac	1.87	—	1.1
Th	1.80	1.76	1.3
Pa	1.63	1.60	1.5
U	1.54	1.51	1.7
Np	1.50	1.47	1.3

[a] 1.1 to 1.2.

Abstracted primarily from Laves, L., in *Theory of Alloy Phases,* American Society for Metals, Metals Park, Ohio, 1956, 131 and Pauling, L., *The Nature of the Chemical Bond,* 3rd ed., Cornell University Press, Ithaca, N.Y., 1960, 93.

where n is the number of valence electrons. The values obtained in this way appear to give more consistent results than those of Pauling when applied to metallic solutions. These data are given in Table 10-4.

No single scale of electronegativity gives results which are in agreement for all different types of solutions or compounds. As would be expected, Pauling's data are best for cases where the bonding is primarily ionic. Gordy's values appear to be more applicable to metallic bonding than do Pauling's. Other electronegativity scales are available.[23]

Table 10-4
GORDY ELECTRONEGATIVITY DATA

Element	r	n	x
Ag	1.53	1	0.91
Al	1.26	3	1.48
As	1.21	5	2.04
Au	1.50	1	0.92
B	0.88	3	1.91
Ba	2.17	2	0.93
Be	1.06	2	1.38
Bi	1.40	5	1.83
C	0.77	4	2.52
Ca	1.78	2	1.03
Cd	1.48	2	1.13
Cr	1.25	6	2.24
Cs	2.25	1	0.78
Cu	1.35	1	0.96
Ga	1.26	3	1.48
Ge	1.22	4	1.77
Hg	1.50	2	1.12
In	1.44	3	1.36
K	1.96	1	0.82
Li	1.34	1	0.96
Mg	1.40	2	1.16
Mn	1.18	7	2.60
Mo	1.36	6	2.09
N	0.74	5	3.01
Na	1.54	1	0.90
Nb	1.43	5	1.76
P	1.10	5	2.19
Pb	1.46	4	1.56
Rb	2.11	1	0.79
S	1.04	6	2.58
Sb	1.41	5	1.82
Sc	1.61	3	1.27
Se	1.17	6	2.35
Si	1.17	4	1.82
Sn	1.40	4	1.61
Sr	1.93	2	0.98
Te	1.37	6	2.08
Ti	1.45	4	1.57
V	1.30	5	1.93
Zn	1.31	2	1.21
Zr	1.58	4	1.48

Note: Also see Reference 28.

After Gordy, W., *Phys. Rev.*, 69, 604, 1946. With permission.

10.2.3. Relative Valence Effect

The relative valence effect also was first proposed by Hume-Rothery as a means for the prediction of the extent of solid solubility. This effect may be expressed as follows: an element of lower valence is more likely to have a greater solubility for an element of higher valence than the opposite case.

Consider, for example, the Cu-Si system. Copper has a valence of 1 and silicon a valence of 4. The solid solubility of Si in Cu is about 14%; this is in contrast to that of Cu in Si which is about 2%. Silicon is covalently bonded and follows the (8-N) rule. This rule gives the number of nearest neighbors, where 8 is the maximum number of s and p levels and N is the column in the periodic table and is equal to four, or more. Thus, each Si ion has four nearest neighbors. Each ion in this lattice shares one of its four electrons with each of its nearest neighbors, thus completing all their outer electron shells by forming covalent bonds. The substitution of a Cu ion for a Si ion in such a lattice disrupts the covalent sharing of the electrons because only one electron is furnished by the Cu atom. The three additional electrons needed to complete the covalent bonding with the nearest neighbors are no longer available. Thus, the presence of Cu causes a situation in which the covalent bonding is incomplete. Increasing amounts of Cu in substitutional solid solution in the Si lattice increase the disruption of the covalent bonding, and increase the energy. It is, thus, expected that the solid solubility of Cu in Si would be limited.

In the case where Cu is the solvent, the bonding is of the metallic type, not covalent. The substitution of each Si ion for a Cu ion adds three electrons over and above those which normally would be present in a lattice composed only of Cu ions. In a half-filled zone like that of Cu, the additional electrons from Si atoms are readily accommodated; many states are available. The electron:atom ratio increases, and consequently, E_F increases also (Equation 5-24, Volume I). Additional Si atoms could be accommodated until, as will be shown later, the Fermi level reaches a limiting point at which, according to the rigid-band model, it reacts with the zone wall. The limit of solid solubility is reached at that concentration.

Normal metals dissolved in other normal metals behave in the same way as noted above. This is shown in Figure 10-2 (also refer to Section 7.9.2 and Figure 7-6, Volume II). It is apparent from Figure 10-2 that the valence of the solute atom, as well as its electronegativity, plays an important part in the determination of the extent of its solid solubility in a given base. Here, the solubility range diminishes as the valence of the solute increases. This occurs because the higher the valence of the solute, the more quickly E_F approaches the limiting condition. This is also shown in Figure 10-4.

In both Figures 10-2 and 10-4, the relative extent of solubility can be roughly approximated by

$$\text{Range Ratio} \propto \frac{1}{Z_B - Z_\alpha} \qquad (10\text{-}4)$$

where Z_α is the valence of the solvent and Z_β is that of the solute element. This is shown in Table 10-5.

The relatively fair agreement between the range ratios and the extent of solid solubility at low temperatures emphasizes the importance of valence. Better agreement would require the incorporation of the effects of other variables (size, electronegativity, etc.) which have been neglected in this oversimplified approach.

The use of integral values for valences simplifies matters. However, as noted in Section 3.11.1 (Volume I), elements in the crystalline state frequently show nonintegral valences. This is particularly true when they are in solid solutions in other metals. Variations in valences may occur depending upon the particular element, the quantity in solution, the kinds and amounts of other elements also in solution, and whether or not ordering (superlattices) takes place.

10.2.3.1. Electron:Atom Ratio

Neglecting other factors, it is apparent from the foregoing that the greater the val-

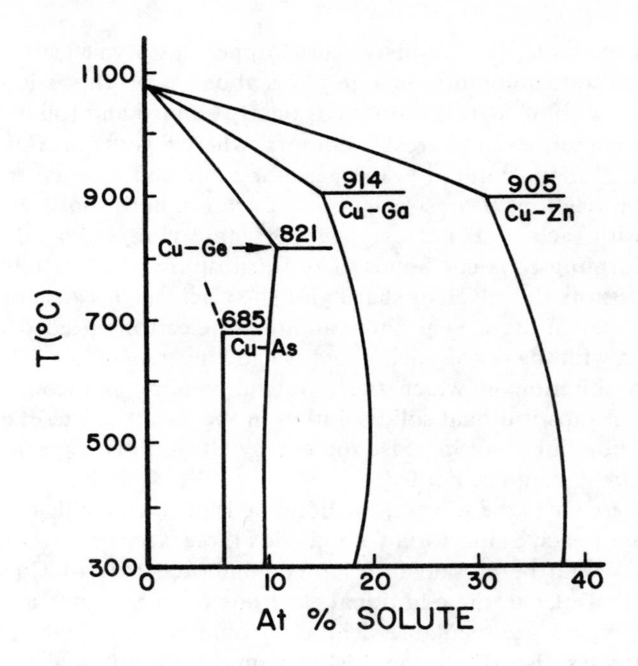

FIGURE 10-4. Extent of some terminal solid solutions in copper. (After Hume-Rothery, W., *The Structure of Metals and Alloys,* Chemical Publishing, New York, 1939, 62. With permission.)

Table 10-5
APPROXIMATE RANGE RATIOS OF SOME SOLID SOLUTIONS

Element	H. -R. valence	Range ratio	Approximate range (at. %)
Ag base	1	—	—
Cd	2	1	42
In	3	½	19
Sn	4	⅓	9
Cu base	1	—	—
Zn	2	1	38
Ga	3	½	19
Ge	4	⅓	10
As	5	¼	7

ence of a solute element, the more limited is its solid-solubility range. Consider the electron:atom ratios of the alloys noted in Table 10-5 at the limits of solubilities as given in Table 10-6.

On the basis of these rough calculations it appears that solid solubility is reached when the average electron:atom ratio is approximately 1.35. If this ratio is regarded as a factor affecting the extent of solid solubility, then it is easy to see why elements with larger valences show smaller limits of solubility than do elements with smaller valences. The solubility depends upon the number of excess electrons which a given solute adds to the base. It will be noted that this is the only variable in the rough approximation given by Equation 10-4. Elements with larger valences contribute more

Table 10-6
ELECTRONOM RATIOS OF ALLOYS AT APPROXIMATE LIMITS OF SOLUBILITIES

Approximate alloy composition	Electron contribution	Electron/ atom (e/a)
42% Cd, 58% Ag	$0.42 \times 2 + 0.58 \times 1 = 0.84 + 0.58$	1.42
19% In, 81% Ag	$0.19 \times 3 + 0.81 \times 1 = 0.57 + 0.81$	1.38
9% Sn, 91% Ag	$0.09 \times 4 + 0.91 \times 1 = 0.36 + 0.91$	1.27
38% Zn, 62% Cu	$0.38 \times 2 + 0.62 \times 1 = 0.76 + 0.62$	1.38
19% Ga, 81% Cu	$0.19 \times 3 + 0.81 \times 1 = 0.57 + 0.81$	1.38
10% Ge, 90% Cu	$0.10 \times 4 + 0.90 \times 1 = 0.40 + 0.90$	1.30
7% As, 93% Cu	$0.07 \times 5 + 0.93 \times 1 = 0.35 + 0.93$	1.28

Note: Using Hume-Rothery Valences; see Table 10-14.

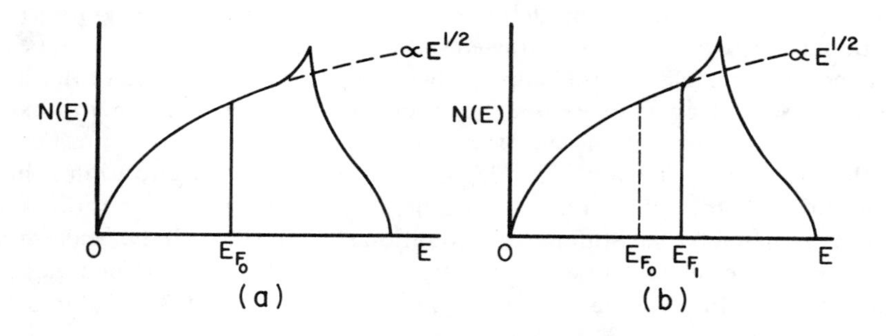

FIGURE 10-5. (a) Fermi level of an unalloyed noble metal; (b) Jones model of the Fermi level at a phase boundary.

excess electrons per atom and approach the limiting electron:atom ratio at lower concentrations than do those of smaller valences.

It previously was indicated that changes in the Fermi level affect the range of solid solubility. It was shown (Equation 5-24, Volume I) that E_F is a function of the electron:atom ratio (see Section 7.9.2, Volume II). This means that E_F must approach a limiting value at which a new phase appears. It was shown previously that each zone can accommodate two electrons per atom (Equation 5-88b, Volume I). An electron:atom ratio of about 1.35 corresponds to a zone which is approximately 67% filled. The Fermi energy of an alloy at the limit of solid solubility, thus, must correspond to this same degree of filling of the zone represented by the given electron/atom ratio. This is shown in Figure 10-5. The half-filled zone of a noble metal is indicated in Figure 10-5(a). Here E_{F_o} corresponds to a half-filled zone and a density of states well below any interaction with the zone boundary in the rigid-band model. And, N(E) obeys the Sommerfeld relationship (Equation 5-21, Volume I). As the electron:atom ratio increases, so does E_F (Equation 5-24, Volume I). The electrons occupying states conforming to Equation 5-21 behave as though they were nearly free. This behavior continues until the curve of N(E) vs. E departs from the Sommerfeld, or nearly free, behavior. This is the point at which $\bar{k} \rightarrow \bar{k}_c$ (Section 5.8.4, Volume I). Here the Fermi energies of the electrons in the zone interact with the zone walls. After this, additional electrons, supplied by an increase in the concentration of the element in solution, cause the values of $\bar{k}$ to become increasingly greater (Figure 5-34, Volume I). In $\bar{k}$ space, the

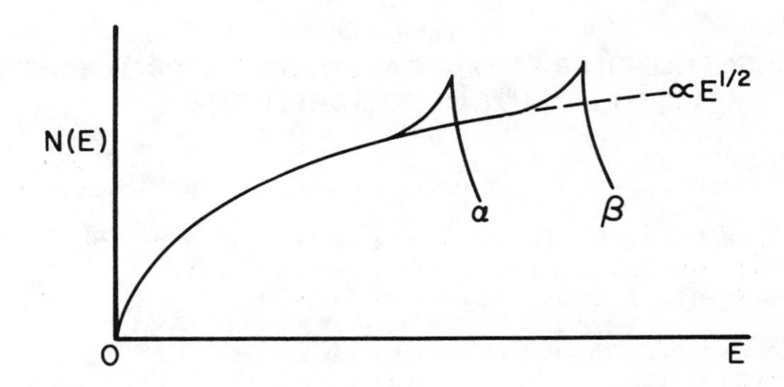

FIGURE 10-6. Curves of densities of states for two possible phases in a given binary alloy system. (Modified from Cottrell, A. M., *Theoretical Structural Metallurgy,* St. Martin's Press, New York, 1957, 136.)

electrons are considered (in this model) to occupy spherical iso-energy contours as long as the Sommerfeld relationship is followed. As a first approximation, it can be considered that the Sommerfeld relationship is obeyed up to the point where the inscribed spherical surface of electron energy, E_F, just touches the zone wall. An approximation is made below to show that the electron:atom ratio responsible for the Fermi level being at this point is approximately 1.35. Beyond the alloy concentration responsible for this, additional amounts of alloy atoms increase the electron:atom ratio and states above those predicted by the Sommerfeld equation start to fill until the peak is reached. Beyond the peak, fewer and fewer states are available at increasing energies. This is equivalent to the filling of the corners of the zone in Figure 5-34 (Volume I). This process continues until all available states are filled.

Now suppose that two phases, α and β, are possible for a given alloy system. The curves of the density of states of both phases are shown, in an exaggerated way, in Figure 10-6. Up to the point at which the N(E) curve of the α phase starts to depart from the Sommerfeld behavior, both phases are possible. Beyond this point the N(E) curve for α starts to peak and the density of states for the α phase is greater than that of the β phase for a given energy increment. Thus, for a given electron:atom ratio, E_F will be lower for α than for β. Here, the α phase will be more probable than the β phase. Beyond the α peak it requires more energy to fill the remaining states in the α zone than it does in the β zone. So, for a given energy increment it is easier to fill states in the β zone, since these still follow the Sommerfeld behavior at this concentration. Thus, for electron densities beyond the α peak, the β phase is the more probable on an energy basis. Therefore, when the electron:atom ratio is such that E_F corresponds to states at, or beyond, the peak of the curve for the α phase, the β phase should be expected to appear; the solid-solubility has been exceeded and the alloy is two-phase: α and β.

The above explanation is based upon nearly free electron behavior, in a rigid-band model with constant energy gaps, up to $\bar{k}_x = \bar{k}_c$, for purposes of simplification. Deviations from spherical energy contours occur in pure, noble metals (see Figure 10-16). In addition, it is implied that the nearly free electron behavior for the β zone is possible and that this is not the case for the almost-filled α zone. It also is assumed that the zone shape and small, constant energy gaps between the zones remain unaffected by alloying. This is the basis for the simplifying approximation that the β phase will appear when the density of states for the α alloys departs from the Sommerfeld behavior. Additional limitations of this approach are given in Section 10.6.6.

It is now clear why the electron:atom ratio reaches a limiting value which is less than that of the maximum zone capacity of two. This also can be shown by considering the energy in terms of $\bar{k}_{max}$ and, in turn, by λ_{min}.

Starting with Equations 5-24 and 3-26 (Volume I)

$$E_{max} = \frac{h^2}{8m}\left[\frac{3N}{\pi V}\right]^{2/3} = \frac{h^2 \bar{k}^2_{max}}{8\pi^2 m} = \frac{h^2}{8\pi^2 m}\left[\frac{2\pi}{\lambda_{min}}\right]^2$$

and simplifying

$$\left[\frac{3N}{\pi V}\right]^{2/3} = \frac{1}{\pi^2} \cdot \frac{4\pi^2}{\lambda^2} = \frac{4}{\lambda^2_{min}}$$

Solving for λ_{min}

$$\lambda_{min} = 2\left[\frac{\pi V}{3N}\right]^{1/3} \qquad (10\text{-}5)$$

The substitutional solid solutions (α phases) of noble metals under consideration here have FCC lattices. In this case, the most widely spaced planes are {111}; these are the planes effectively limiting the zone (see Figure 5-29a and Section 5.8.3.2, Volume I). Reflections, for incident radiation perpendicular to these planes, will first occur where

$$\lambda_{min} = 2d = \frac{2a}{[h^2 + k^2 + \ell^2]^{1/2}} = \frac{2a}{\sqrt{3}} \qquad (10\text{-}6)$$

The FCC unit cell contains four atoms and has a volume of a^3. Thus the volume per atom is

$$\frac{V}{N_o} = \frac{a^3}{4} \qquad (10\text{-}7)$$

where N_o is the number of atoms per unit cell. This is reexpressed to give the lattice parameter as

$$a = \left[\frac{4V}{N_o}\right]^{1/3} \qquad (10\text{-}8)$$

Equation 10-8 is substituted into Equation 10-6 to obtain

$$\lambda_{min} = \frac{2}{\sqrt{3}}\left[\frac{4V}{N_o}\right]^{1/3} \qquad (10\text{-}9)$$

Equations 10-5 and 10-9 are equated

$$2\left[\frac{\pi V}{3N}\right]^{1/3} = \frac{2}{\sqrt{3}}\left[\frac{4V}{N_o}\right]^{1/3} \qquad (10\text{-}10)$$

Table 10-7
DATA FOR SOME ISOMORPHOUS SYSTEMS

Element	Radius CN 12	Electronegativity x	Valence Z	Lattice type
Cu	1.28	1.9	1	FCC
Ni	1.24	1.8	0.6	FCC
Au	1.44	2.4	1	FCC
Ag	1.44	1.9	1	FCC
Pd	1.37	2.2	0.6	FCC

After simplifying and cubing,

$$\frac{N}{N_o} = \frac{\pi\sqrt{3}}{4} = 1.36 \qquad (10\text{-}11)$$

Thus the electron:atom ratio computed in this way compares well with the data in Table 10-6.

The extent of the terminal solid solubility of ternary solid solution alloys may be estimated in this way also. Approximations made for such alloys usually are less accurate than those for binary alloys (see Section 10.6.6.1).

10.2.4. Lattice Type

Other factors being equal, extended ranges of solid solubility will be favored when the crystal structures of the pure solvent and solute are alike. The Cu-Ni, Au-Ag, and Ni-Pd systems are good examples of this type of behavior. Data for these elements are given in Table 10-7.

The Cu and Ni atoms crystallize in the same lattice type and have small differences between their radii, electronegativities, and valences. This set of conditions is ideal for solid solubility and leads to their isomorphous (complete solid solubility) constitutional behavior. The Au-Ag pair shows similar behavior, except for their electronegativity difference; the difference of 0.5 units might lead to the expectation of more limited solubility than in the Cu-Ni case. Despite this, Au and Ag also form an isomorphous system. The elements Ni and Pd show the greatest size disparity (about 10%) of all the pairs, the other factors being favorable. This, too, might be considered to have an effect upon solid solubility. Again, complete solid solubility is manifested. Observations of this type emphasize the empirical nature of the Hume-Rothery rules.

If the components of a binary system have different lattice types, it is probable that an isomorphous system will not form. This observation results from the idea that the crystal structures of the alloys are likely to change as the concentration varies across the constitution diagram from one component to the other. In other words, some transition between the two crystal structures may be expected. If this occurs, it may involve the presence of one or more intermediate crystal types. This prevents continuous solid solubility and places limits upon the extent of the terminal solid solutions. Another accommodation between the two crystal types may consist of a mixture of the two terminal solid solutions.

10.3. SOLVUS CURVES

The most commonly observed solvus curves show decreasing solid solubility with decreasing temperature. Other binary systems show opposite behavior where the solid

Table 10-8
DATA FOR ELEMENTS SHOWN IN FIGURE 10-4

Atom	H-R Valence Z_β	Radius CN12	Gordy x	Isotherm (°C)
Cu	(1)	1.28	0.96	(1083)
Zn	2	1.37	1.21	905
Ga	3	1.39	1.48	914
Ge	4	1.39	1.77	821
As	5	1.48	2.04	685

solubility of the terminal solutions diminishes with increasing temperature. Still others show intermediate behavior (Figures 10-2 and 10-4). This behavior results from the nature and properties of the solid solution and of the precipitating phase, since the solvus represents the equilibrium between these two phases.

The extent of solid solubility of an alloying element is influenced strongly by the value of its electronegativity as compared to that of the host element (Section 10.2.2). The greater the difference between these values, the greater the tendency to form compounds. This may occur even when the size and other factors are favorable for solutions. In general, the more stable the intermediate phase, the more limited is the extent of the primary solid solution and the steeper is the slope of the solvus. Most of the alloy systems which form stable intermetallics have solvus curves which indicate decreasing solid solubilities of the terminal solid solutions with decreasing temperatures. Precipitation will occur upon cooling alloys with compositions which are near that of the solvus at elevated temperatures. The properties of such alloys will depend upon the nature of the precipitate, its precipitation mode and the degree to which it is dispersed. For example, an alloy which forms a continuous network of a brittle, intermediate compound at the grain boundaries, will demonstrate significantly lower mechanical properties than one in which the precipitate is finely divided and well dispersed within the grains as well as in the grain boundaries. The physical properties of alloys also may be affected by precipitates as described in Section 9.14.2 (Volume II). It will be shown later (Section 10.6.6) that many intermetallic phases correspond to certain electron:atom ratios.

The previously noted parameters, size, relative valence, as well as electronegativity all influence the behavior of the solvus. While the shapes of these curves cannot be predicted from current theory, when they are plotted as functions of the electron:atom ratio, many do show similar behavior. The observed variations often can be accounted for individually on the basis of lattice distortion and/or the character of the precipitating phase.

10.4. SOLIDUS CURVES

As a first approximation, the lowering of the solidus curves is approximately a function of the valence difference between the solute and solvent atoms, the size factor, and the electronegativity of the atoms involved. Data for the elements shown in Figure 10-4 are given in Table 10-8.

It will be noted that Zn and Ga have very similar radii and, even though their electronegativities and valences are different, both show about the same lowering of the solidus. The element Ge has nearly the same radius as these elements, but its electronegativity and valence differences are greater. It might have been expected that the lowering of the solidus by Ga should have been intermediate between those induced

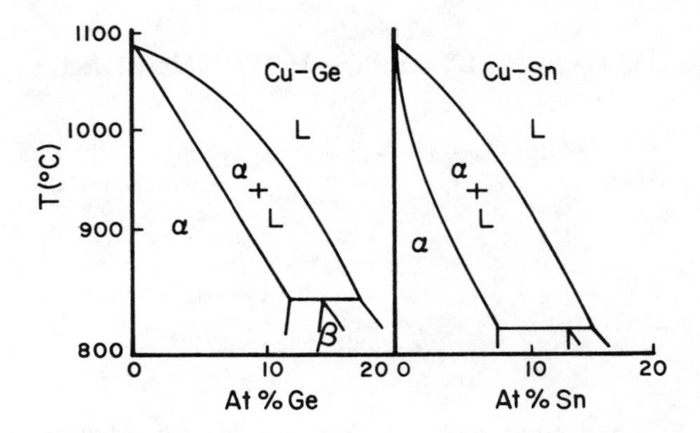

FIGURE 10-7. Portions of the Cu-Ge and Cu-Sn phase diagrams. (After Hume-Rothery, W., *The Structure of Metals and Alloys,* Chemical Publishing, New York, 1939, 71. With permission.)

by Zn and Ge. As, with the greatest set of parametric differences, shows the greatest depression. The reason for the slightly anomalous behavior of Ga is not known.

The Cu-Sn and the Cu-Ge systems give an indication of the effect of the size factor upon the slope of the solidus. The elements Ge and Sn have the same valence and about the same electronegativity. The difference between the diameters of Cu and Ge is about 9%, while that between Cu and Sn is about 15%. On the basis of the size factor, the range of solubility for Ge in Cu should be larger than that of Sn in Cu. Thus, it would be expected that the slope of the solidus of the Cu-Sn system should be steeper than that of the Cu-Ge system. This is verified by the portions of the phase diagrams for these elements shown in Figure 10-7.

Similar behavior also occurs in the Mg-In and Mn-Al systems. Here Mg and In have compatible size factors, but that of Al is marginal. Where relatively large size disparities occur, such as those of the Cu-Sn and Mg-Al systems, the solidus has a steeper slope and the two-phase, α + L, field is wider than in the more compatible cases.

In general, the greater the disparity between the size factors, the steeper is the solidus curve. This is responsible for the increased extent of the α + L field. The greatest depressions of the solidus occur in cases in which the limit of solid solubility is negligibly small.

10.5. LIQUIDUS CURVES

The valences of the solute atoms play a dominant part in the behavior of liquidus curves. This is shown in Figure 10-8. Here, in both Cu- and Ag-base binary alloys, the slopes of the liquidus curves increase as the valences of the solute elements increase. This parallels the behavior noted in the discussions of the roles of relative valences and electronom ratios (Sections 10.2.3 and 10.2.3.1). When the data for these alloy systems are plotted as a function of the electronom ratio, the liquidus curves fall within a narrow band. The result is shown in Figure 10-9.

The behavior indicated in the figure shows that the depression of the liquidus is a regular function of the equivalent composition. This is a function of solute valence; for example: 20% Zn, 13-1/3% Ga, 10% Ge, and 8% As are equivalent compositions. The product of the atom percent of each of the elements and its respective valence is 40. Thus, each equivalent composition represents a single electronom ratio. Therefore, the independent variable in Figure 10-9 actually provides a comparison of the various

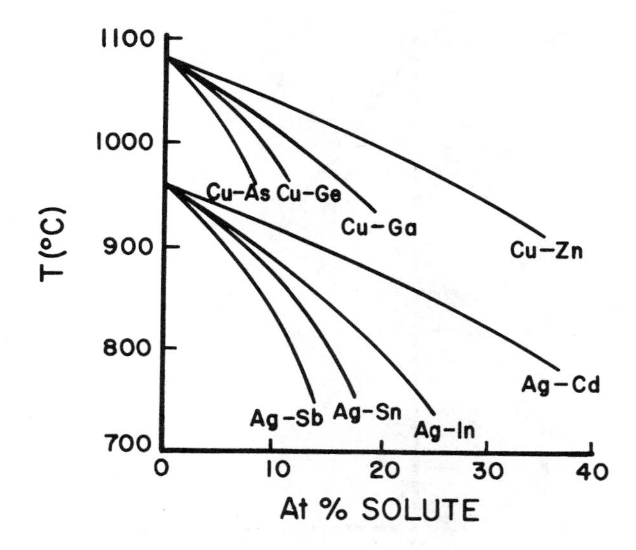

FIGURE 10-8. Liquidus curves for some Cu- and Ag-base binary alloy systems. (After Hume-Rothery, W., *The Structure of Metals and Alloys,* Chemical Publishing, New York, 1939, 62. With permission.)

liquidus data on the basis of common electronom ratios for all of the alloys. This also is the equivalent of making the comparison at common Fermi levels (Equation 5-24, Volume I).

In alloy systems in which the solvent and solute atoms belong to different periods in the periodic table, the interactions may be more complex. However, where the size factors are correct, similar depressions of the liquidus are found. Some examples of this are given by dilute solutions of the Cu-Ge, Cu-Si, Ag-Sb, Ag-Bi, Ag-Al, and Ag-In systems. In cases where relatively large electronegativity differences exist, and the tendency toward compound formation is strong, the electron/atom ratio alone cannot explain the observed liquidus curves.

Where only small differences exist in size factors and valence, the effect of the solute element upon the liquidus should be small. In the Ag-Au system, the presence of small amounts of Ag has a very small influence upon the liquidus (Figure 10-10a). Similar behavior is shown by Cu in the Cu-Ni system. The phase equilibria of Ag-Pd, Au-Pd, and Cd-Mg systems also show this. This behavior could be explained on the basis that the atoms concerned are of nearly equal sizes and valences. Within such limitations, a minimum of disruption is caused by the addition of more solute atoms.

Where the atomic size factors differ by more than about 8%, relatively large changes in the liquidus can occur. This is shown, for example, by the Cu-Pt system, as in Figure 10-10b. Binary phase diagrams of other noble metal-transition metal combinations also show this. Some transition metal-transition metal combinations, as for example, Ni-Pt, have liquidus curves similar to that shown in Figure 10-10b.

Another type is one in which the liquidus shows a minimum as a function of composition (Figure 10-10c). This is usually observed between pairs of transition elements. Some systems showing this include: Cr-Fe, Cu-Mn, Fe-Pd, Fe-V, Ni-Pd, and Pt-Mn. However, other systems such as K-Rb, Cu-Au, and As-Sb have liquidus curves of similar shapes. The appearance of the minimum in the liquidus represents intermediate behavior approximately between that of an isomorphous system and a eutectic system with terminal solid solutions.

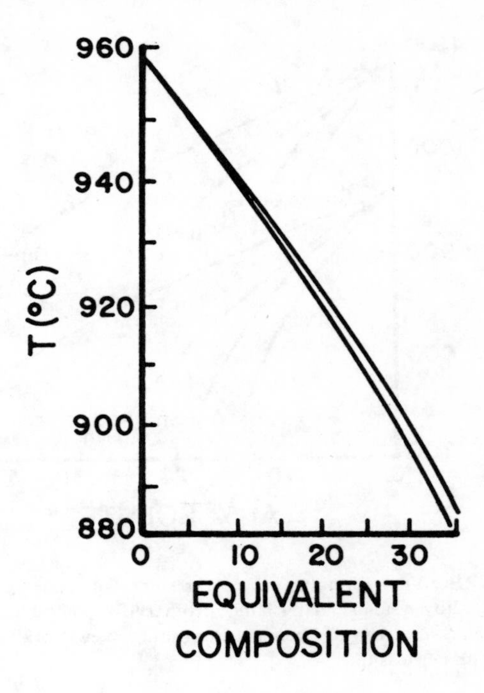

FIGURE 10-9. Liquidus curves of silver-base alloys as a function of equivalent composition. (After Hume-Rothery, W., *The Structure of Metals and Alloys,* Chemical Publishing, New York, 1939, 64. With permission.)

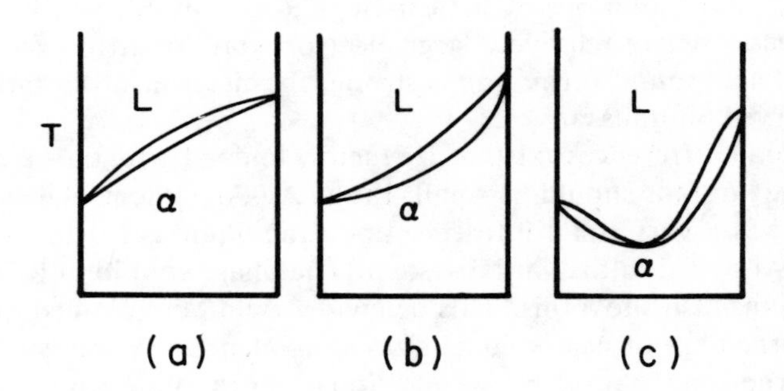

FIGURE 10-10. Liquidus and solidus curves of some binary isomorphous systems. (After Hume-Rothery, W., *The Structure of Metals and Alloys,* Chemical Publishing, New York, 1939, 67. With permission.)

10.6. INTERMEDIATE PHASES

It is helpful to describe briefly the forces acting between assemblies of the atoms before dealing with intermetallic, or intermediate, phases. The principle bond types found in intermediate phases are ionic, covalent, and metallic types. These are discussed in greater detail in subsequent sections.

In ionic crystals, oppositely charged ions are bound together by electrostatic attraction. This arises from the transfer of outer electrons from one atom, leaving it positively charged, to another atom which then becomes a negative ion. Common rock salt, NaCl, is a frequently used example of this behavior. The Na atom gives up its single valence electron and becomes a positively charged ion. The Cl atom takes up this electron to complete its $3p^6$ level and becomes a negative ion. The equilibrium positions between such ions in a lattice result primarily from a balance between the attractive and repulsive (coulombic) forces (see Section 10.6.1).

Covalent bonding occurs when the valence electrons are shared in pairs between atoms; this effectively completes the outer bands of each atom. Each such pair of electrons, each one of which has opposite spin, bonds two adjacent atoms. Such electron pairs have a high probability of being found in the region between the two ion cores. As would be expected, crystals bonded in this way have highly directional properties (Figure 10-17a). Such bonding usually exists where the valence of one of the ions is 4, or greater, and the electronom ratio is 4. The high bond strength arises from the bond hybridization and neutralization of the unbalanced charges on the nuclei which results from the sharing of the electrons. These compounds follow the (8-N) rule (Section 10.2.3). In many cases, the bonding is not homogeneous, but may be partially ionic and partly covalent. Other bonding may consist of mixed ionic, covalent, and metallic bonding (see Sections 10.6.7 and 10.6.8).

The metallic bond is best pictured as one in which a valence electron is considered as being shared by all of the ions in the crystal. This behavior has been considered as an extremely dilute case of unsaturated, covalent bonding.

It was previously shown (Figure 5-16, Volume I) that the greatest probability of finding an electron in a hydrogen molecule was between the ions. Now consider the lithium atom. Its ground state configuration is $1s^2\ 2s$. Its 1s states are tightly bound, so, to a first approximation, it can be regarded as being similar to hydrogen. When two Li atoms are brought together, a situation like that described earlier for hydrogen should occur. When a third Li atom is brought near the original pair, a significant difference is manifested. Many more vacant electron states are available to the Li atoms as increasing numbers of them are assembled (Section 3.7 and Figure 3-4, Volume I). In the case of hydrogen, the next available states are much higher in energy. The 2s electron from the third Li atom will have its spin the same as one of the other two electrons. This can be accommodated by one of the two additional spin states available to the Li atoms. The electron from the third atom can resonate between its original ion and the other two. As more and more Li atoms are assembled, more and more electrons are "shared" in this way. These valence electrons resonate among the ions independently of one another and, on the average, the number of electrons associated with a given ion will be just enough to keep it neutral electrically.

Lithium crystallizes in the BCC lattice. It, thus, has eight nearest neighbors. Assume that a bond is formed between each ion and each of its nearest neighbors, as in Figure 10-11.

Each dashed line joining the body-centered ion is considered to be a bond. Each ion will have eight such bonds. Each bond arises between two ions and, since the bonds cannot be counted twice, there will be four bonds per ion. Each of the atoms has one valence electron available. Thus, there will be 1/4 of an electron available for each bond. The electron can resonate between ions in the crystal in a manner not unlike that of a covalent bond. The normal covalent bond requires the sharing of two electrons of opposite spins. The metallic bond is, obviously unsaturated, since only 1/4 of the number of electrons required for normal covalent bonding is available. This is insufficient to complete the outer bands as in covalent bonding. Under these conditions, the electrons do not "belong" to any given ions, but resonate at random between all of the ions in the lattice.

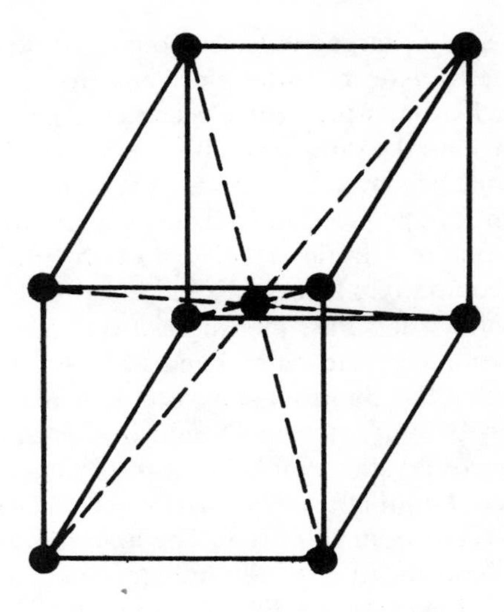

FIGURE 10-11. Bonds between nearest neighbors in a BCC lattice.

This behavior illustrates the concept of the nearly free electron. As a rule of thumb, metallic bonding occurs when the electronom ratio is less than 4.

A rough approximation of the cohesive energy of normal metals may be obtained from simplified models of their band structures. Metals belonging to group I of the Periodic Table have partially filled valence bands so that their valence electrons show nearly free behavior. The valence bands of those elements belonging to this group, which have small, principle quantum numbers, may be considered to be unaffected by overlap with other bands. This permits calculations based upon the association of one valence electron with one ion. The cohesive energy, E_c, is measured by the energy required to raise an electron from within the partly filled valence band up to a level at least to the top of the band (Figures 5-19 and 5-20, Volume I) or, out of the well (Figure 3-3d, Volume I). The calculation is simplified further by assuming that the band width is constant for all $\overline{k}$.

The lowest state in the valence band is considered to have no kinetic energy; it is the state of maximum potential energy. This is the level to which any electron kinetic energy is added. The kinetic energy is the repulsion component and arises from the Pauli Exclusion Principle. Considering that the filled portion of the band is occupied by nearly free electrons, the potential and kinetic energies are added to obtain the total energy. In the case of metals in which the valence band is approximately half-filled, E_F will be close to the center of the band. The average kinetic energy is $\overline{E} = 3/5E_F$ (Equation 5-25, Volume I), and is positive in sign.

In a band which is approximately half-filled, the energy range from the center to the top of the band is about equal to E_F. Since $\overline{E}$ is $3/5E_F$, when measured from the bottom of the band, the electron with average energy will lie at about $(E_F - 2/5E_F)$ from the center of the band. Therefore, the energy required to raise such an electron to the top of the band must be equal and opposite to

$$-E_c \simeq -E_F + (E_F - 2/5E_F) = -2/5E_F$$

This is in terms of the potential energy which must be overcome for such excitation. It will be recalled (Section 5.7, Volume I) that the energy of atoms is lowered when they are brought together[1,2] (calculations made using data from Table 5-2, Volume I).

Metal	E_c(eV/atom)
Li	$0.35E_F$
Na	$0.35E_F$
K	$0.42E_F$
Rb	$0.46E_F$
Cs	$0.55E_F$
Cu	$0.50E_F$
Ag	$0.55E_F$
Au	$0.73E_F$

The values observed for Ag, Cs, and Au are appreciably higher than the approximation because the principle quantum numbers of their s electrons (5 and 6) are high. Here, the ratios of the ionic diameters to the interionic distances approach unity as n increases. The electrons in the closed shells of adjacent ions can mutually interact under these conditions. The bonding is increased by this. If the cohesive energy increment produced by this effect is estimated as contributing an additional $1/5E_F$ to the bond strength, the cohesive energy may be approximated as being about $3/5E_F$. This is in closer agreement with the data for Ag, Cs, and Au than the previous approximation.

It should be noted that these rough approximations neglect the potential of the ion core under consideration and those of its neighbors. This amounts to the omission of crystal structure from these approximations. In addition, the repulsive interactions of all but the one electron being examined also are neglected.

Wigner and Seitz provided a more complete approximation for E_c by considering the solid to be made up of proximity cells (atomic polyhedra) the volumes of which were used to approximate spheres of given radii, r (see Section 5.8.2, Volume I). This allowed the use of the Schrodinger time-independent equation (Equation 3-34, Volume I) with appropriate values of V(r) instead of V(x). The resultant wave function for Na becomes very flat at about half the interionic distance, thus permitting the valence electrons as being considered to be nearly free. This allowed the use of the same expression for the kinetic energy as was used above. The cohesive energy, then, was calculated from the negative sum of the potential, kinetic, and ionization energies per ion.

The cohesive energies of most normal metals lie within about 2 to 4 eV/atom. Those of the transition elements, not included in either of the approximations given above, range from about 3 to 9 eV/atom.

10.6.1. Bonding Energy of Ionic Crystals

The bonding energy for a pair of ions such as Na^+ and Cl^- is given by

$$U = -\frac{e^2}{r} + \frac{\beta}{r^s} \tag{10-12}$$

where e is the charge, r is the distance between them, and s and β are constants for a given substance. The first term of Equation 10-12 is the attractive energy, the second is the repulsive energy. The bonding energy of a mole is

$$U = N\left[-\frac{e^2}{r} + \frac{\beta}{r^s}\right] \tag{10-12a}$$

in which N is Avogadro's number. The coefficient of N is unity since although there are 2N ions in a mole, the number of ion pairs is 2N/2.

If ϕ is the electrostatic potential at a lattice point resulting from all the ions except the one being considered, then

$$\phi = -\frac{Ze\alpha_m}{r} \tag{10-13}$$

where Z is the number of bonds, or valence, and α_m is the Madelung constant. It accounts for the coulombic interactions of all other ions with a given ion and is

$$\alpha_m = \sum_i \frac{\pm 1}{r_i} \tag{10-14}$$

where r is the distance between nearest neighbors and r_i is the distance between the ion being considered and the ith ion. The Madelung constant is about 1.75 for the NaCl structure, but varies between 1.63 and 25.0 depending upon the type of ionic lattice formed. Multiplying both sides of Equation 10-13 by NZe

$$NZe\phi = \frac{N(Ze)^2\alpha_m}{r} \tag{10-15}$$

gives the attractive energy per mole. When the ions are brought close together, the filled shells cause them to act like hard spheres. The repulsive term in Equation 10-12 is operative only at short distances and results from the resistance of the ions to overlapping caused by exclusion; s is known as the Born exponent. The total energy of an ion pair is

$$U_T(r) = -\frac{\alpha_m(Ze)^2}{r} + \beta r^{-s} \tag{10-16}$$

The variation of the total energy with the distance between the ions is

$$\frac{dU_T(r)}{dr} = \frac{\alpha_m(Ze)^2}{r^2} - s\beta r^{-s-1}$$

The energy is minimized at equilibrium and

$$\frac{\alpha_m(Ze)^2}{r^2} = s\beta r^{-s-1} \tag{10-17}$$

Multiplying both sides of Equation 10-17 by r

$$\frac{\alpha_m(Ze)^2}{r} = s\beta r^{-s}$$

Then,

$$\beta r^{-s} = \frac{\alpha_m(Ze)^2}{rs} \tag{10-18}$$

This is substituted into Equation 10-16 to give

$$U_T(r) = \frac{-\alpha_m(Ze)^2}{r} + \frac{\alpha_m(Ze)^2}{rs}$$

or, for a pair of ions,

$$U_T(r) = \frac{-\alpha_m(Ze)^2}{r}\left(1 - \frac{1}{s}\right) \qquad (10\text{-}19a)$$

The energy per mole of ions is

$$U_T(r) = -\frac{N\alpha_m(Ze)^2}{r}\left(1 - \frac{1}{s}\right) \qquad (10\text{-}19b)$$

The negative coefficient is the Madelung energy when r is taken as the closest approach of the ions. For alkali halides $U_T(r)$ is about 6 to 8 eV. The Born exponent, s, varies from 5 to 12 (see Table 10-10).

10.6.2. Compressibility of Ionic Crystals

The compressibility is defined, at absolute zero, as

$$K = -\frac{1}{V}\frac{\partial V}{\partial P} \qquad (10\text{-}20)$$

where V is the volume and P is the pressure. The bulk modulus is defined as the reciprocal of the compressibility. From the first law of thermodynamics,

$$\partial U_T(r) \equiv \partial U = -P\partial V \qquad (10\text{-}21)$$

From this

$$\frac{\partial^2 U}{\partial V^2} = -\frac{\partial P}{\partial V} \qquad (10\text{-}22)$$

Equation 10-20 may be reexpressed, by means of Equation 10-22, as

$$\frac{1}{KV} = -\frac{\partial P}{\partial V} = -\left[-\frac{\partial^2 U}{\partial V^2}\right] = \frac{\partial^2 U}{\partial V^2} \qquad (10\text{-}23)$$

An expression can be obtained for the right-hand side of Equation 10-23. This is found by starting with

$$\frac{\partial U}{\partial V} = \frac{\partial r}{\partial V} \cdot \frac{\partial U}{\partial r} \qquad (10\text{-}24)$$

For the NaCl type lattice the unit cell length, a, is twice the closest approach of the ions, r_o, or

$$a = 2r_o$$

The volume occupied per ion pair is calculated from the number of ion pairs in a unit cell:

corner atoms:	$1/8 \times 8$	=	1 atom(s)
face atoms:	$1/2 \times 6$	=	3
edge atoms:	$1/4 \times 12$	=	3
body atom:	1×1	=	1/8 atoms = 4 pairs

So, the volume per ion pair, V_I, in terms of hard spheres, is

$$V_I = \frac{a^3}{4} = \frac{(2r_o)^3}{4}$$

and the volume of a mole is

$$V = \frac{8Nr_o^3}{4} = 2Nr_o^3 \qquad (10\text{-}25)$$

This, of course, will be different for other types of ionic lattices. Equation 10-25 can be differentiated with respect to r to give

$$\frac{\partial V}{\partial r} = 6Nr_o^2$$

and inverted to obtain

$$\frac{\partial r}{\partial V} = \frac{1}{6Nr_o^2} \qquad (10\text{-}26)$$

which can be used in Equation 10-24 as

$$\frac{\partial U}{\partial V} = \frac{1}{6Nr_o^2} \cdot \frac{\partial U}{\partial r_o} \qquad (10\text{-}27)$$

Equation 10-27 is now differentiated with respect to V

$$\frac{\partial^2 U}{\partial V^2} = \frac{1}{6Nr_o^2}\frac{\partial}{\partial V}\frac{\partial U}{\partial r_o} = \frac{1}{6Nr_o^2} \cdot \frac{\partial}{\partial V}\left[\frac{\partial U}{\partial V} \cdot \frac{\partial V}{\partial r_o}\right]$$

$$\frac{\partial^2 U}{\partial V^2} = \frac{1}{6Nr_o^2}\left[\frac{\partial U}{\partial V\,\partial r_o}\right] = \frac{1}{6Nr_o^2}\left[\frac{\partial}{\partial r_o} \cdot \frac{\partial U}{\partial r} \cdot \frac{\partial r_o}{\partial V}\right] \qquad (10\text{-}28)$$

Substituting Equation 10-26 in Equation 10-28

$$\frac{\partial^2 U}{\partial V^2} = \frac{1}{6Nr_o^2}\left\{\frac{\partial}{\partial r_o}\left[\frac{\partial U}{\partial r} \cdot \frac{1}{6Nr_o^2}\right]\right\} \qquad (10\text{-}29)$$

Or,

$$\frac{\partial^2 U}{\partial V^2} = \frac{1}{36N^2r_o^2} \cdot \frac{\partial}{\partial r_o}\left[\frac{1}{r_o^2}\frac{\partial U}{\partial r}\right] \qquad (10\text{-}30)$$

Equation 10-16 may be written as

$$U_T(r) \equiv U = N\ [\beta r_o^{-s} - \alpha_m Z^2 e^2 r_o^{-1}] \qquad (10\text{-}31)$$

Upon differentiation with respect to r_o, Equation 10-31 becomes

$$\frac{\partial U_T(r)}{\partial r_o} = N\left[-s\beta r_o^{-s-1} + \alpha_m Z^2 e^2 r_o^{-2}\right] \qquad (10\text{-}32)$$

Now, dividing both sides of Equation 10-32 by r_o^2

$$\frac{1}{r_o^2} \cdot \frac{\partial U_T(r)}{\partial r_o} = N\left[-s\beta r_o^{-s-3} + \alpha_m Z^2 e^2 r_o^{-4}\right] \qquad (10\text{-}33)$$

Equation 10-33 can be substituted into Equation 10-30 to give

$$\frac{\partial^2 U}{\partial V^2} = \frac{N}{36N^2 r_o^2} \frac{\partial}{\partial r_o}\left[-s\beta r_o^{-s-3} + \alpha_m Z^2 e^2 r_o^{-4}\right] \qquad (10\text{-}34)$$

When the indicated differentiation is performed

$$\frac{\partial^2 U}{\partial V^2} = \frac{1}{36N\, r_o^2}\left[s(s+3)\beta r_o^{-s-4} - 4\alpha_m Z^2 e^2 r_o^{-5}\right] \qquad (10\text{-}35)$$

The condition for minimum energy is found by equating Equation 10-32 to zero to obtain

$$\frac{s\beta}{r^{s+1}} = \frac{\alpha_m Z^2 e^2}{r_o^2} \qquad (10\text{-}36)$$

Dividing both sides of Equation 10-36 by r^3

$$\frac{s\beta}{r^{s+4}} = \frac{\alpha_m Z^2 e^2}{r_o^5} \qquad (10\text{-}37)$$

This is now substituted into Equation 10-35 to give

$$\frac{\partial^2 U}{\partial V^2} = \frac{1}{36N\, r_o^2}\left[(s+3)\frac{\alpha_m Z^2 e^2}{r_o^5} - \frac{4\alpha_m Z^2 e^2}{r_o^5}\right]$$

$$\frac{\partial^2 U}{\partial V^2} = \frac{1}{36N\, r_o^2}\left[(s+3) - 4\right]\frac{\alpha_m Z^2 e^2}{r_o^5}$$

$$\frac{\partial^2 U}{\partial V^2} = \frac{\alpha_m Z^2 e^2}{36N\, r_o^7}(s-1) \qquad (10\text{-}38)$$

Now, returning to Equation 10-23 and using Equation 10-38

$$\frac{1}{K} = V\frac{\partial^2 U}{\partial V^2} = V\frac{\alpha_m Z^2 e^2 (s-1)}{36N\, r_o^7} \qquad (10\text{-}39)$$

Then, using Equation 10-25, Equation 10-39 becomes

$$\frac{1}{K} = 2Nr_o^3 \cdot \frac{\alpha_m Z^2 e^2}{36N\, r_o^7}(s-1)$$

or,

$$\frac{1}{K} = \frac{\alpha_m Z^2 e^2 (s - 1)}{18 r_o^4} \qquad (10\text{-}40)$$

The results given here are for NaCl type lattices. Variations in lattice types must be taken into account when applying these equations to other ionic lattice types by making the appropriate changes in Equation 10-25.

The Born exponent is found from Equation 10-40 to be

$$s = 1 + \frac{18 r_o^4}{\alpha_m Z^2 e^2 K} \qquad (10\text{-}41)$$

And from Equation 10-36,

$$\beta = \frac{\alpha_m Z^2 e^2}{s\, r_o^{1-s}} \qquad (10\text{-}42)$$

See Table 10-10.

10.6.3. Ionic Solids in General

It will be noted from Equation 10-19b that most of the energy results from the coulomb attraction of the ions. The repulsion contribution is quite small. In addition, rather large variations in s result in correspondingly small variations in the bonding energy of the crystal. Some experimental values of s are given in Table 10-9, after extrapolation to the absolute zero of temperature.

The value of s can be approximated from the electron configuration of the ions involved (Table 10-10). This approximation is made by taking the average of the values of the two ion cores of interest.

The smaller the distance of closest approach of the ions, the stronger will be the attractive bonding energy between them. However, as the ions become closer, the short-range repulsion energies become operative (Equation 10-12). These become appreciable as the outer levels of the ions start to overlap. Some distortion of the ions occurs. A minimum in the energy is observed (Figure 10-12). This is usually approximated as being the point at which the hard spheres just touch each other, and is shown at r_o in the figure.

Since the equilibrium between the forces of attraction and repulsion actually occurs after some ionic overlap has taken place, the ions must "penetrate" each other. This penetration can have significant effects. If the outer electrons of the negative ion are loosely bound to it, these can be shared covalently with adjacent positive ions. This adds a component of covalent bonding to the predominant ionic (coulombic) bonding. Thus, the bonding energy is not purely ionic, but is mixed and is increased by the covalent component. One rule of thumb is that additional covalent bonding may take place when the ionization potential of the metal ion is at least 2eV greater than that of the nonmetal. In any event, it appears that some effects of covalency probably are present and that the bonding is not purely ionic. This is shown in Table 10-11. Here, most of the experimental values are slightly larger than the calculated values. At least a part of the difference could be ascribed to a covalency effect, especially where the differences are relatively large.

An estimate of the bonding energy of NaCl-type lattices can be made based on Equation 10-19b. Use is made of an average value of s = 9 and covalent effects are neglected. Then,

Table 10-9
EXPERIMENTAL VALUES FOR THE BORN EXPONENTS

Crystal	s
LiF	5.9
LiCl	8.0
LiBr	8.7
NaCl	9.1
NaBr	9.5

After Slater, J. C., *Phys. Rev.*, 23, 488, 1924. With permission.

Table 10-10
THE BORN EXPONENT

Fundamental ion core	Ion core structure	Born exponent, s
He	$1s^2$	5
Ne	$He + 2s^2\,2p^6$	7
Ar	$Ne + 3s^2\,3p^6$	9
Kr	$Ar + 3d^{10}\,4s^2\,4p^6$	10
Xe	$Kr + 4d^{10}\,5s^2\,5p^6$	12

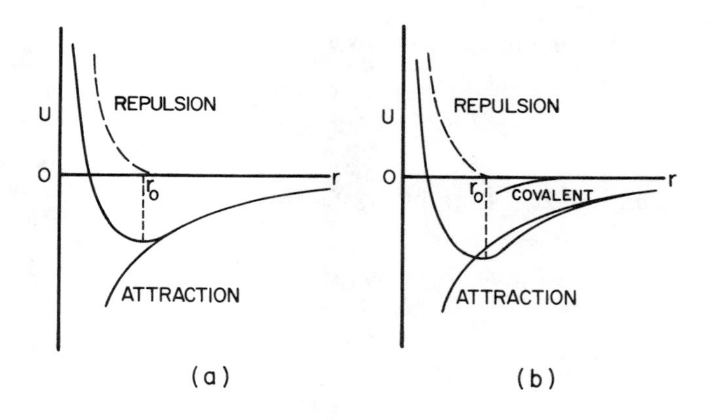

FIGURE 10-12. Ionic bonding energy as a function of distance between two oppositely charged ions. (a) Coulombic attraction and repulsion; (b) Additional effect of covalent contribution. (After Mott, N. F. and Gurney, R. W., *Electronic Processes in Ionic Crystals*, 2nd ed., Dover, New York, 1948, 9.)

$$U_T(r) = -\frac{8}{9}\,\frac{N\alpha_m(Ze)^2}{r} \qquad (10\text{-}19c)$$

An expression for the distance of closest approach, r, can be obtained in terms of the molecular weight, M, and the density, ϱ, of the crystal:

Table 10-11
BONDING ENERGIES OF CRYSTALS WITH NaCl STRUCTURES

	Bonding energy (kcal/mol)			Bonding energy (kcal/mol)	
Compound	Experimental	Calculated	Compound	Experimental	Calculated
LiF	−242.3	−242.2	KF	−189.8	−189.1
LiCl	−198.9	−192.9	KCl	−165.8	−161.6
LiBr	−189.8	−181.0	KBr	−158.5	−154.5
LiI	−177.7	−166.1	KI	−149.9	−144.5
NaF	−214.4	−215.2	RbF	−181.4	−180.4
NaCl	−182.6	−178.6	RbCl	−159.3	−155.4
NaBr	−173.6	−169.2	RbBr	−152.6	−148.3
NaI	−163.2	−156.6	RbI	−144.9	−139.6

Abstracted from Kittel, C., *Introduction to Solid State Physics,* John Wiley & Sons, New York, 1966, 98.

$$\rho = \frac{4(M^+ + M^-)}{Na^3} = \frac{4M}{Na^3}$$

Rearranging,

$$\frac{1}{a^3} = \frac{N\rho}{4M}; \quad \frac{1}{a} = \left[\frac{N\rho}{4M}\right]^{1/3}$$

The lattice parameter, a, is approximated by $a \simeq 2r_o$. Thus,

$$\frac{1}{r_o} \simeq 2\left[\frac{N\rho}{4M}\right]^{1/3}$$

Using this expression and Z = 1 for this crystal type gives

$$U_T \simeq -\frac{8}{9} N\alpha_m e^2 \cdot 2\left[\frac{N\rho}{4M}\right]^{1/3}$$

Collecting terms gives

$$U_T \simeq -\frac{16}{9} N^{4/3}\alpha_m e^2 \left[\frac{\rho}{4M}\right]^{1/3} \qquad (10\text{-}19d)$$

For $\alpha_m = 1.75$,

$$U_T \simeq -1.78(0.602 \times 10^{24})^{4/3} \times 1.75(4.8 \times 10^{-10})^2 \frac{1}{1.59}\left[\frac{\rho}{M}\right]^{1/3}$$

$$U_T \simeq -1.78 \times 0.508 \times 10^{32} \times 1.75 \times 23.04 \times 10^{-20} \times 0.628 \left[\frac{\rho}{M}\right]^{1/3}$$

$$U_T \simeq -22.90 \times 10^{12} \left[\frac{\rho}{M}\right]^{1/3} \text{ (erg/mol)}$$

$$U_T \simeq -\frac{22.90 \times 10^{12}}{10^7}\left[\frac{\rho}{M}\right]^{1/3} = -22.90 \times 10^5 \left[\frac{\rho}{M}\right]^{1/3} \text{(J/mol)}$$

$$U_T \simeq -22.90 \times 10^5 \times 0.24 \left[\frac{\rho}{M}\right]^{1/3} = -5.50 \times 10^5 \left[\frac{\rho}{M}\right]^{1/3} \text{(cal/mol)}$$

NaCl is used to illustrate this. Using a density of 2.2 g/cm³ and a molecular weight of 58.5, $U_T \simeq -179$ kcal/mol. This is in good agreement with the values given in Table 10-11.

It will be recognized that other types of ionic crystals require that suitable changes be made in the equation given above for the density.[24]

10.6.4. Ionic Lattice Types

Ionic compounds of the M^+X^- type crystallize in one of the lattices typified by the cesium chloride, sodium chloride, or zinc sulfide (zinc-blend) structures. These are shown in Figure 10-13, based on the hard-sphere concept.

All of these structures maintain the ratio M/X = 1 to maintain their compositions and electrical balances. The CsCl structure, like the BCC, contains one atom at the center of the unit cell and 8 at its corners, CN8. It, thus, maintains the 1:1 ratio in a BCC-like lattice. The NaCl structure, described in the previous section, has CN6. The ZnS lattice is comparable to a FCC lattice which contains four additional internal atoms of the opposite kind. It too preserves the 1:1 ratio, but with CN4.

It is interesting to note that Goldschmidt employed data from crystals such as these to obtain values for the radii of ions. By comparing the lattice parameters of compounds of a given metal and various halogens, a nearly constant difference is obtained which can be associated with the radius of the given metal ion, based upon a hard-sphere model. This approach will be employed to examine the three ionic lattice types, remembering that the atomic radius is not a constant for a given element under all conditions.

Consider the CsCl lattice (Figure 10-14). From this it can be seen, where r_1 and r_2 are the respective radii of the smaller and larger ions, that

$$[2(r_1 + r_2)]^2 = a^2 + 2a^2 = 3a^2$$

from which

$$a = \frac{2}{\sqrt{3}}(r_1 + r_2)$$

Usually, the radius of M^+ is smaller than that of the halogen ion. If the radius of the nonmetallic ion is increased, a point will be reached at which the center ion will no longer touch all the corner ions. This is the case when $a > 2r_2$. Then

$$2r_2 \geqslant \frac{2}{\sqrt{3}}(r_1 + r_2)$$

or

$$\sqrt{3}\, r_2 \geqslant r_1 + r_2$$

$$0.73\, r_2 \geqslant r_1$$

or

$$r_2 \geqslant 1.37 r_1$$

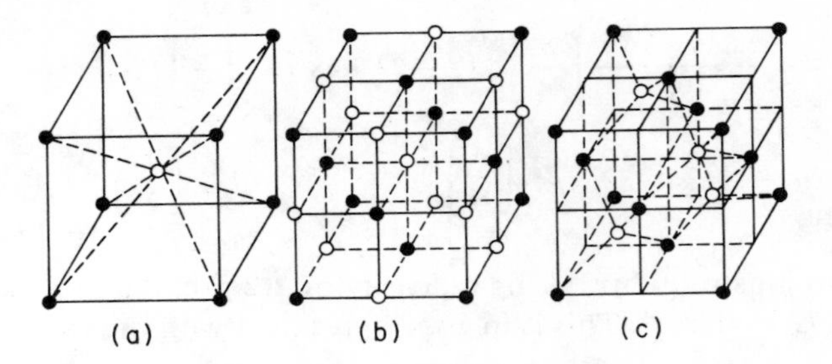

FIGURE 10-13. Typical ionic crystal lattices: (a) CsCl, (b) NaCl, and (c) ZnS.

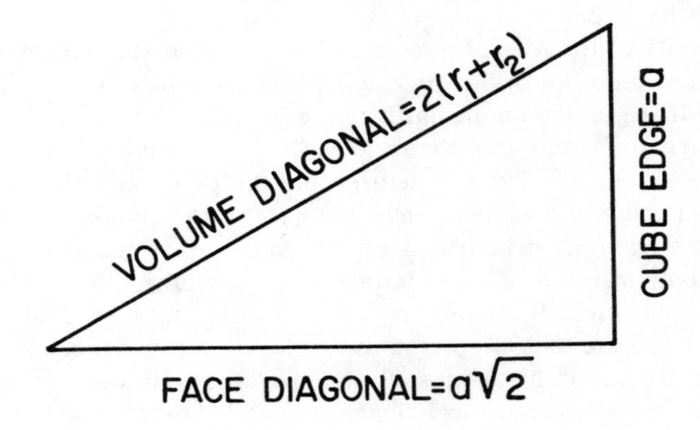

FIGURE 10-14. Geometry of the CsCl-type structure based upon hard-sphere packing.

If the ratios of the ionic diameters exceed this value, the bonding energy would no longer be a minimum and some other lattice structure should be favored which would minimize the energy.

Now consider the NaCl type lattice. This is less dense and has a lower coordination number than the CsCl lattice. The lattice parameter of this lattice is given by

$$a = 2(r_1 + r_2)$$

The square of the diagonal of a face is $2a^2$. Thus,

$$(\text{diagonal})^2 = 2 \times 4(r_1 + r_2)^2$$

and

$$\text{diagonal} = 2\sqrt{2}(r_1 + r_2)$$

The diagonal equals $4r_2$

$$4r_2 = 2\sqrt{2}\,(r_1 + r_2)$$

or

$$2r_2 = \sqrt{2}\,(r_1 + r_2)$$

Table 10-12
SOME EXAMPLES OF TYPES OF IONIC CRYSTALS

CsCl	NaCl	ZnS
LaZn	MgSe	BeS
CeZn	CaSr	BeTi
CeCd	SrSe	AlP
LaHg	BaSe	AlS
CeHg	MnSe	CdSe
BaZn	PbSe	ZnSe
LaHg	SnTe	InP
BaCd	MnTe	GeSb

Thus, when $r_2 > 2.39r_1$, the ions in the NaCl lattice will no longer touch; the bonding energy will not be a minimum and another type of lattice should be favored.

Similarly, the ZnS structure should not be favored if $r_2 > 4.55\ r_1$.

In summary, the CsCl structure should be favored when the ratios of the ionic radii lie between 1 and 1.37, the NaCl structure for ratios between 1.37 and 2.39, and the ZnS lattice for ratios between 2.39 and 4.55. These ratios, at best, are crude approximations since the ions are not hard spheres; ionic radii are not always constant (Section 10.2.1) and some ionic distortion occurs when the ions are at their equilibrium lattice positions. This uncertainty is shown by some metal-halogen salts with the CsCl lattice which have ratios of ionic radii of about 1.5; others having the ZnS structure have ratios which vary from 1.9 to 5.6.

It also should be noted, on the basis of the space and symmetry concepts (Sections 10.2.1.2 and 10.2.1.3), that the packing becomes less efficient and the symmetry diminishes as the lattice types change from CsCl to NaCl to ZnS.

Intermetallic compounds crystallizing in these lattice types are predominantly ionic in character. Their bonding energies are mostly due to the Madelung contribution (Equation 10-19b). Some examples of the three classes of structures are given in Table 10-12.

10.6.5. Laves Phases

Unlike the ionic lattice types just discussed, Laves phases are more densely packed than FCC or HCP lattices. These phases also are explained by means of the hard-sphere model. In other words, the relative sizes of the ions are critical in achieving this density of packing. This kind of packing occurs when the ratio of the diameters of the ions is 1.22 (ideally 1.225). Such phases have stoichiometric compositions of the type AB_2 and the bonding is metallic in character. The three lattice types of this class of compounds are represented by $MgCu_2$, $MgZn_2$, and $MgNi_2$.

The ideal crystalline array is one in which each A ion has 12B and 4A nearest neighbors, CN16, and each B ion has 12. These are like interpenetrating lattices in which like ions touch one another, but the unlike ions do not touch. This gives an average CN13.3. The cubic $CuMg_2$ lattice is of this form. The relationship between the hexagonal $MgZn_2$ lattice and the cubic $MgCu_2$ lattice is analogous to the relationship between the HCP and FCC lattices. The hexagonal $MgNi_2$ type seems to be intermediate in character between the $MgCu_2$ and $MgZn_2$ lattices; it has been considered to be a "mixture" of the two types.

The electron:atom ratio appears to be of secondary importance in the determination of which lattice type will be formed, when the ionic sizes are correct. The $MgCu_2$ lattice forms when the electron:atom ratio lies between 1.33 and 1.75; $MgZn_2$ and $MgNi_2$ when the electron:atom ratios lie in the ranges 1.81 to 2.05 and 1.78 to 1.90, respec-

Table 10-13
SOME EXAMPLES OF LAVES PHASE TYPES

$MgCu_2$	$MgZn_2$	$MgNi_2$
$AgBe_2$	$BaMg_2$	$NbZn_2$
$CuAl_2$	KPb_2	$ScFe_2$
$NaAu_2$	$MnBe_2$	$ThMg_2$
$PbAu_2$	$MoFe_2$	UPt_2
$TaCo_2$	$TiZn_2$	β-Co_2Ti

Table 10-14
SOME ELECTRON PHASES

Electron: Atom Ratio[2]

3:2		21:13	7:4
β Phase (BCC)	β Manganese (complex-cubic)	γ Phase (complex-cubic)	ε Phase (HCP)
AgCd	Ag_3Al	Ag_5Cd_8	$AgCd_3$
AgMg	Au_3Al	Ag_5Hg_8	Ag_3Sn
AgZn	Cu_5Si	Ag_5Zn_8	Ag_5Al_3
AuCd	$CoZn_3$	Au_5Cd_8	Ag_5In_3
AuMg		Au_5Zn_8	$AgZn_3$
AuZn		Cu_5Cd_8	$AuCd_3$
CuBe		Cu_5Hg_8	$AuZn_3$
CuZn		Cu_5Zn_8	Au_3Sn
Cu_3Al		Cu_9Al_4	Au_3Hg
Cu_3Ga		Cu_9Ga_4	Au_5Al_3
Cu_5Sn		Cu_9In_4	$CuCd_3$
CoAl		$Cu_{31}Si_8$	$CuZn_3$
FeAl		$Cu_{31}Sn_8$	Cu_3Ge
NiAl		Co_5Zn_{21}	Cu_3Si
		Fe_5Zn_{21}	Cu_3Sn
		Ni_5Zn_{21}	$FeZn_7$

[a] Hume-Rothery Valences:

Group I: Cu, Ag, Au	1	Group IV: Si, Ge, Sn, Pb	4
Group II: Be, Mg, Zn, Cd, Hg	2	Group V: P, As, Sb, Bi	5
Group III: Al, Ga, In	3	Transition elements: Fe, Co, Ni, Pd, etc.	0*

*Note:** It is incorrect to consider that ions with zero valences do not engage in bonding. In many cases, this condition arises from behavior which indicates that they accommodate as many electrons in vacant d states as they contribute to bonding. This can result in mixed metallic and ionic bonding such as in NiAl (Section 10.6.9.).

Abstracted from Barrett, C. S., *Structure of Metals,* McGraw-Hill, New York, 1952, 237. With permission.

tively. These concentrations appear to be those at which the Fermi energies of the electrons in the zone begin to interact with the various zone walls.

The size constraints account for the fact that most compounds of this type have had compositions very close to that of AB_2; they usually do not exist over wider ranges of composition. It is apparent that the lattice strain energy would be increased greatly for other compositions, since some ions would be required to occupy incorrect lattice sites. This severely limits the composition ranges of this class of phases. Thus, the geometrical assembly of ions of the correct sizes for these lattice types determines the compositions and lattice types of this class of phases rather than the electron:atom ratio.

The bonding of intermetallic compounds crystallizing in these forms is largely metallic in character. Some examples of these compounds are given in Table 10-13.

10.6.6. Electron Phases

Hume-Rothery and others have shown that many of the constitution diagrams of the noble metals with metals belonging to the B subgroups show very similar phase relationships and structures. The Cu-Zn diagram is typical of these. Starting with the noble metal, these systems show the presence of FCC α solid solutions, β phases which have disordered BCC lattices, γ phases which can have up to 52 atoms in a complex-cubic unit cell, high-temperature δ phases which are complex-cubic with voids at some lattice sites, HCP ε phases, and finally, η HCP terminal solid solutions. The first three of these (α,β,γ) will be given primary consideration. These phases are not chemical compounds because they are stable over wide ranges of compositions.

It was shown (Section 10.2.3.1 and Table 10-6) that the α solutions range up to compositions corresponding to average electron:atom ratios of about 1.35. Similar empirical observations showed that the β, γ and ε phases exist at compositions corresponding to the approximate electron:atom ratios of 3/2, 21/13, and 7/4, respectively (Table 10-14).

These phases do not show the close stoichiometric relationships required of the Laves phases; they usually exist over wide ranges of compositions. The ranges in composition have been explained by means of the analogy that the alloys with the electron:atom ratios given above act as solvents and form the equivalent of substitutional solid solutions which extend on each side of the specified compositions. This is not unreasonable since the ions in question generally fall within the 15% size-factor limitation. The composition ranges of some of these electron phases are shown in Figure 10-15.

These phases have metallic bonding, conduct electricity, and have, as would be expected, low bonding energies, usually of the order of 5 kcal/mol, or less.

The original theory attempting to explain these phases was presented by H. Jones. It involves the calculation of the electron:atom ratio at which the Fermi energy surface of the electrons in a zone just touches the zone wall, as in Section 10.2.3.1. This approximation assumes that the electrons are nearly free, that the band gaps between the zones are small as well as constant, and that the shape of the zone is not changed by alloying. It follows from this that the energy surface contours will be spherical and that E_F will become tangent to the zone walls at points. Then, the compositions corresponding to the electron:atom ratios at which the inscribed, spherical Fermi energy surface just becomes tangent to the zone walls will be at the limit of the stable phases (see Figure 10-6). This method is used for the calculation of the electron:atom ratios for the α, β and γ phases in the following sections. The ε phase, with an electron:ion ratio of 7:4, is in general agreement with the Jones zone (Figure 5-29 and Equation 5-88b, Volume I) when the ions are considered to be hard spheres and the assumptions

Table 10-15
HEATS OF FORMATION (KCAL/ MOL AT 298 K) OF PHASES WITH NiAs STRUCTURES[a]

Phase	−ΔH	Phase	−ΔH
FeS	11.4 ± 0.2	NiAs	6.5[b]
NiS	11.1 ± 0.7	CoSe	5.0 ± 1.3
CoS	11.0 ± 0.6[c]	CoTe	4.5 ± 1.3
FeSe	9.0 ± 0.5	CoSb	5.0 ± 0.4
NiSb	7.9 ± 1.0[d]	NiSe	5.0 ± 1.5
Ni_3Sn_2	7.5 ± 1.0[d]	NiTe	4.5 ± 1.5
MnAs	6.9[e]	CoSn	3.6 ± 0.3[d]
MnSb	6.0[e]	AuSn	3.6 ± 0.05[f]
Co_3Sn_2	2.7 ± 0.2[c]	MnBi	2.4

[a] All data are taken from Kubaschewski, O. and Evans, E. Ll., *Metallurgical Thermochemistry,* 3rd ed., Pergamon Press, Elmsford, N.Y., 1958, except as noted.

[b] Estimated.

[c] Raynor, G. V., *The Physical Chemistry of Metallic Solutions and Intermetallic Compounds,* N.P.L. Symp. No. 9, 1, 3A, H.M.S.O., London, 1959.

[d] Hultgren, R., Orr, R. L., Anderson, P. D., and Kelley, K. K., *Selected Values of Thermodynamic Properties of Metals and Alloys,* John Wiley & Sons, New York, 1963.

[e] Shchukarev, S. A., Morozova, M. P., and Stolyarova, T. A., *Zh. Obshch. Kim.,* 31, 1773, 1961, quoted after *Chem. Abstr.,* 55, 23,026b, 1961.

[f] Misra, S., Howlett, B. W., and Bever, M. B., *Trans. Met. Soc. AIME,* 233, 749, 1965.

Abstracted from Robinson, P. M. and Bever, M. B., in *Intermetallic Compounds,* Westbrook, J. H., Ed., John Wiley & Sons, New York, 1967, 48. With permission.

given above are made. As noted below and in Section 10.2.3.1, this approach is a simplified one.

More recent data show that the Fermi surface of a half-filled zone of a pure, noble metal touches the zone boundaries, as shown in Figure 10-16 and that the gaps between the zones vary with the wave vector (Section 5.8.4, Volume I). The absence of a spherical Fermi surface only introduces a small error when the Jones spherical approximation is used. It may be applied in all cases except those in which the Fermi surfaces are badly deformed. When this is the case, the nearly free electron approach probably is not valid. It is important to note that the assumption of the appearance of a new phase at the electronom ratio which causes the spherical Fermi surface to become tangent to the zone wall is not necessarily the critical case. The extent of the energy gaps or overlaps between the zones determine this. However, the electronom ratio has a significant influence on the range of terminal solid solutions (Section 10.2.3.1) and affects the structures and compositional ranges of intermetallic compounds. The Jones approach is presented in the following sections, with these limitations in mind, to provide an uncomplicated insight into some of the theory of alloy phases.

10.6.6.1. Alpha Phases

The α phases, or FCC terminal solid solutions, were treated in Section 10.2.3.1 on the basis of X-ray reflections. The Jones approach, which makes use of the reciprocal lattice, is as shown below.

The ionic volume is given by Equation 10-7. In reciprocal space, the volume per ion in the zone is

$$V_Z = \frac{4}{a^3}$$

The radius of the largest sphere which can be inscribed between the {111} planes (see Figure 5-29a, Volume I) is given by Equation 10-6, or in reciprocal space,

$$r = \frac{1}{2d} = \frac{\sqrt{3}}{2a}$$

and the volume of the inscribed Fermi sphere is

$$V_s = \frac{4}{3}\pi\frac{3\sqrt{3}}{8a^3}$$

The ratio of the volume of the inscribed Fermi sphere to the volume of the zone, V_Z, provides the basis for the approximation of the electron:ion ratio. Thus,

$$\frac{V_s}{V_Z} = \frac{\frac{4}{3}\pi\frac{3\sqrt{3}}{8a^3}}{\frac{4}{a^3}} = \frac{\pi\sqrt{3}}{8}$$

But, each zone has the capacity to accommodate two electrons per atom, or

$$\frac{V_s}{V_Z} = \frac{2\pi\sqrt{3}}{8} = 1.36 \text{ electrons/ion}$$

This is in good agreement with the average of the data in Table 10-6 and with the earlier calculation given by Equation 10-11.

10.6.6.2. Beta Phases

The β phases have disordered BCC lattics. (The ordered β phases, which are not considered here, crystallize in CsCl type lattices.) The volume of an atom in the zone in reciprocal space is given by

$$V_Z = \frac{2}{a^3}$$

since the BCC lattice contains two atoms per unit cell. The most widely spaced planes in this lattice are {110} (see Figure 5-29b and Section 5.8.3.3, Volume I). The radius of the Fermi sphere which just touches these planes is

$$r = \frac{1}{2d} = \frac{\sqrt{2}}{2a}$$

and the volume of the inscribed Fermi sphere is

$$V_S = \frac{4}{3}\pi\frac{2\sqrt{2}}{8a^3}$$

When the capacity of the zone to accommodate electrons is taken into account:

$$2\frac{V_S}{V_Z} = 2\frac{\frac{4}{3}\pi\frac{2\sqrt{2}}{8a^3}}{\frac{2}{a^3}} = \frac{\pi\sqrt{2}}{3} = 1.48 \text{ electrons/ion}$$

This also is in good agreement with the empirical ratio of 3/2 found by Hume-Rothery.

It should be stated again that this high degree of agreement between the electron:ion ratios calculated in this way and those determined empirically probably is not entirely warranted for the reasons given in Sections 10.2.3.1 and 10.6.6.

10.6.6.3. Gamma Phases

The lattice of the gamma phase is complex cubic with up to 52 ions per unit cell. It is based primarily upon X-ray reflections, rather than structure, as is the case for the α and β phases and may contain vacant lattice sites. The volume of an ion in this zone, in reciprocal space, is given by

$$V_Z = \frac{52}{a^3}$$

The strongest reflections are observed from {411} and {330} planes. Half the distance between these planes is

$$r = \frac{1}{2d} = \frac{3\sqrt{2}}{2a}$$

The volume of the inscribed Fermi sphere is

$$V_s = \frac{4}{3}\pi\left[\frac{3\sqrt{2}}{2a}\right]^3 = \frac{9\pi\sqrt{2}}{a^3}$$

Again, accounting for the capacity of the zone to accept electrons

$$2\frac{V_S}{V_Z} = \frac{2 \times 9\pi\sqrt{2}}{a^3} \times \frac{a^3}{52} = \frac{9\pi\sqrt{2}}{26} = 1.53 \text{ electrons/ion}$$

This value is slightly lower than the experimentally observed value of $21/13 = 1.62$. The gamma phases have compositional ranges corresponding to electron:ion ratios of from about 1.52 to 1.76. Considering all of the previously noted simplifications, and the facts that this unit cell is of low symmetry and does not approach close packing, this degree of agreement also is not entirely justifiable.

It must be noted that however simplified this approach may be, the Jones reasoning has become a basis for more advanced models of phase relationsbips.

10.6.7. Covalently Bonded Phases

The simplest form of covalent bonding is illustrated by the hydrogen molecule (Figure 5-15, Volume I). This shows the bonding between the two hydrogen atoms which is caused by the overlapping and interaction of the wave functions. The result is a mutual sharing of the electrons from each atom so that the 1s levels of each are filled. The decrease in energy which results from this is shown by the minimum of $E_s + E_R$ in Figure 5-17 (Volume I). Decreases in energy of this kind are responsible for the strength and stability of covalent bonds.

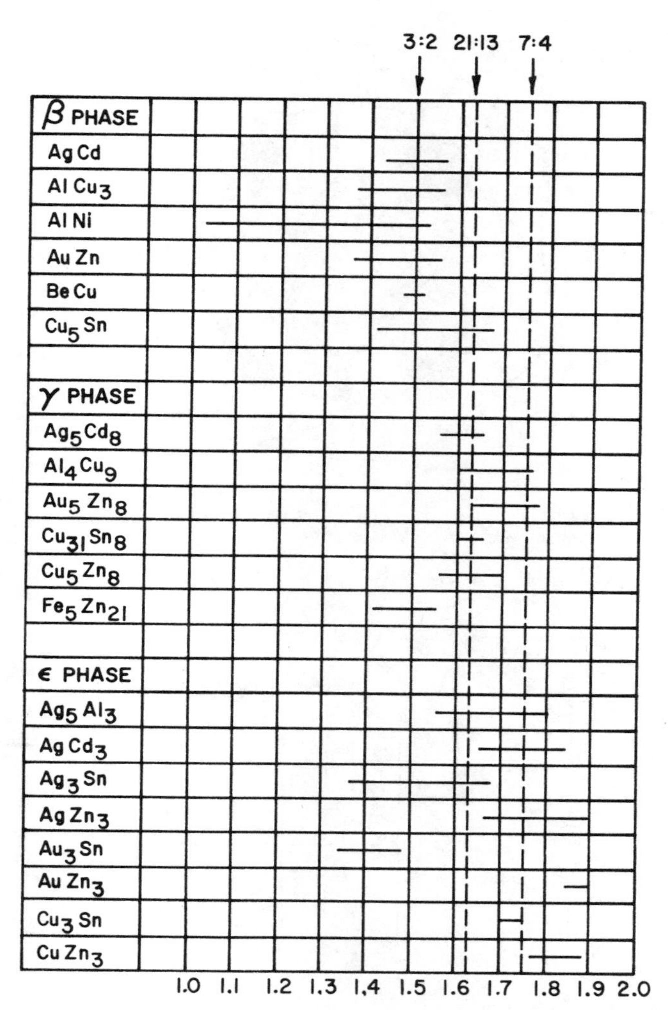

FIGURE 10-15. Ranges of composition of some electron phases. (After Barrett, C. S., *Structure of Metals*, McGraw-Hill, New York, 1952, 238. With permission.)

A more complex situation exists for the case of carbon ($1s^2\ 2s^22p^2$). Here, two unpaired 2p states exist. The first excited state is $1s^22s^12p^3$; it represents a small increase in energy over the ground state. In this case, four unpaired states are present and are known as sp^3 hybridized states. The decrease in energy which occurs when these electrons form bonds is lower than that for the ground state. The hybridization, or combination, of the mixed single s and the three p states results in four equivalent bonds. Bond pairs of this type, also known as tetrahedral hybridization, form equal angles with each other. This high degree of directionality results in the formation of tetrahedra (Figure 10-17a). These are the components of the diamond-cubic lattice (Figure 10-17b), which contains eight of these. Such sharing also conforms with the (8-N) rule previously noted in Sections 10.2.3 and 10.6.

Trigonal hybridization (sp^2) accounts for the crystal structure of graphite and explains the diamagnetic properties of this substance which were described in Section

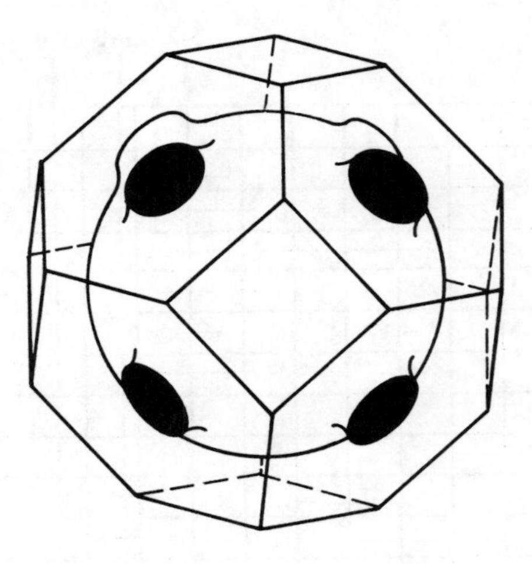

FIGURE 10-16. Fermi surface and Brillouin zone of a noble metal.

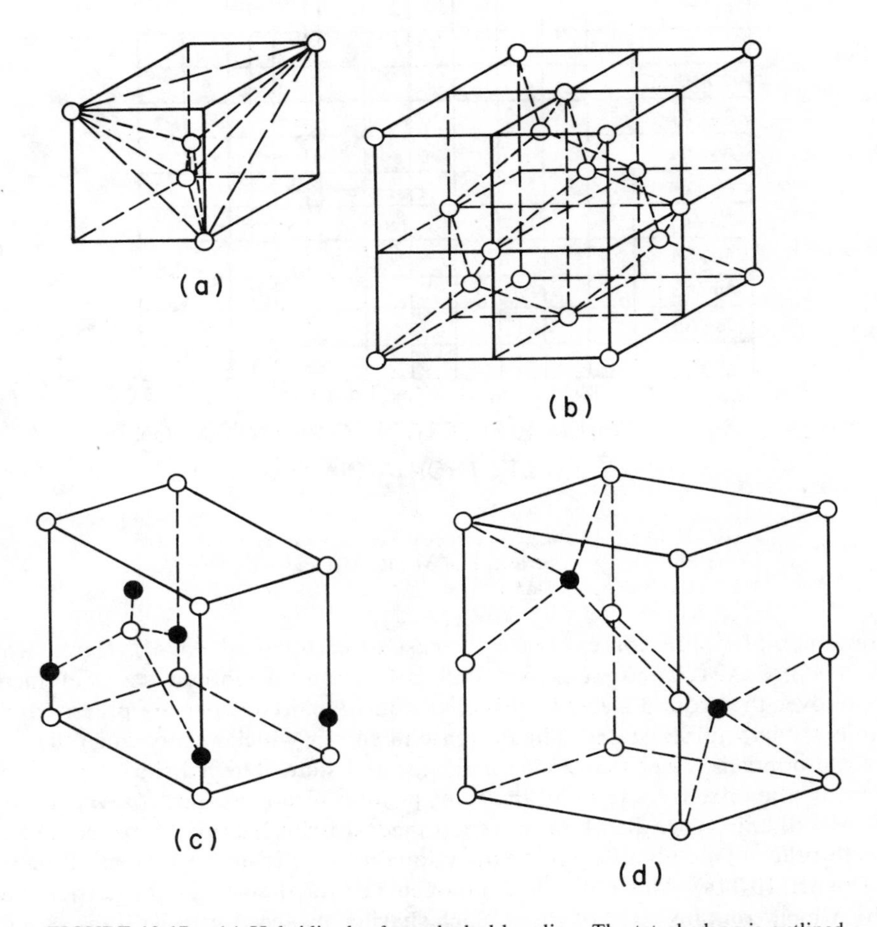

FIGURE 10-17. (a) Hybridized sp^3 tetrahedral bonding. The tetrahedron is outlined by large broken-line segments. (b) The diamond cubic structure (note its similarity to the ZnS structure, Figure 10-13). (c) The wurtzite unit cell. (d) The NiAs unit cell (note that the structure cells of wurtzite and NiAs are hexagonal.)

8.2.1.2 (Volume II). This configuration also explains the lubricating properties of graphite because the sp^2 bonds in the hexagonal, basal plane are strong; but the remaining bond between these planes is weak and permits very easy motion of the basal planes. Other variations in the extent of hybridization can explain the large numbers of structures of organic molecules.

Silicon, germanium, and tin also have outer s^2p^2 electron configurations like that of carbon in the ground state; they also form comparable, hybridized tetrahedral bonds. (Gray tin, the nonmetallic allotrope, shows this behavior below the transformation temperature which occurs at 13°C.) These elements crystallize in the diamond-cubic lattice for the same reasons given for carbon. In all of these cases the four bonding electrons of a given ion are shared in pairs with those of its four nearest neighbors. Thus, the valence states of each ion are completely filled. As noted previously, shared bonds of this kind are strong and stable. In terms of the band model (Figure 5-20, Volume I), since the valence bands are filled, these elements are expected to be either insulators or semiconductors, depending upon the width of the energy gap between the filled valence and empty conduction bands. As shown in Chapter 11, these gaps are such that intrinsic semiconduction is observed.

Comparable behavior is shown by compounds of the form AB, in which the elements belong to groups III and V of the Periodic Table. Generally, the two ions involved are considerably different in size. Their electron:ion ratios are four and they have four nearest neighbors. These compounds form lattices similar to the diamond-cubic in which one of the metal ions occupies the central position as shown in Figure 10-17a. The nonmetal ions constitute an FCC sublattice. Such a composite lattice is known as the zinc-blende structure. As is the case for the elements noted above, the valence bands of these compounds are completely filled at 0 K and they behave similarly at very low temperatures. However, many compounds of this type (also known as III-V compounds), such as GaAs and InSb, have been the subject of much research for semiconductor applications.

Other elements form tetrahedral structures closely related to the zinc-blende lattice. These crystallize in the wurtzite lattice structure (Figure 10-17a). In this case, the nonmetal ions form a HCP sublattice. As is the case for the previous structures, four ions of one kind are symmetrically located about one ion of the other component. Two examples of compounds with this structure are CdSe and β-ZnS. Both the wurtzite and the zinc-blende structures sometimes are called adamantine structures, because they are similar to diamond in luster and hardness.

The lattices of many compounds of these types have defect structures in which some lattice sites normally occupied by metal ions are vacant. No such defects occur upon sites occupied by nonmetal ions. Adjacent lattice sites will have less than four nearest neighbors when defects are present. Some of the hybridized states, thus, cannot enter into bonding. When this occurs, the electron:atom ratio will be a function of the number of vacant sites (see Section 10.6.9).

In addition to the III-V compounds represented by zinc blende and wurtzite, II-VI and I-VII compounds with similar properties also have been investigated for semiconductor applications.

10.6.8. Mixed Bonding

As indicated in Section 10.6.3 and Figure 10-12b, ionic crystals may contain some degree of covalency in their bonding. Varying degrees of mixed ionic and covalent bonding may exist in other crystals. The degree of ionic bonding in such cases may be estimated on the basis of a semi-empirical relationship derived by Pauling (Figure 10-18). Here, the degree of ionic character of the bonding of a two-component compound

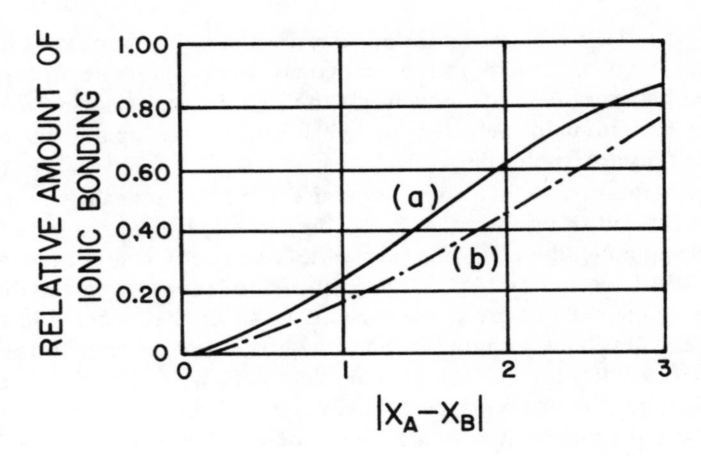

FIGURE 10-18. Relative amounts of ionic bonding as functions of electronegativity differences. [(a) After Pauling, L., *The Nature of the Chemical Bond,* Cornell University Press, Ithaca, N.Y., 1960, 99.) (b) After Hannay, N. B. and Smith, C. P., *J. Am. Chem. Soc.*, 68, 171, 1946.)

is roughly proportional to the absolute difference between the electronegativities of the components. Hannay and Smith give the following relationship (Figure 10-18, curve b):

$$\text{Percent ionic character} = 16 \mid x_A - x_B \mid + 3.5 \mid x_A - x_B \mid^2$$

Where the bonding takes on more covalent or more metallic character, the reliability of predicting the ionic component of bonding decreases. The ionic radii determine the density of the packing and, consequently, the lattice type. Generally, as the density of packing increases, the heat of formation of the compound increases. This, in turn, affects the electronegativity and limits the utility of this parameter as a measure of bond type. However, if a reasonable value of the heat of formation of a compound is given by Equation 10-1, it is safe to consider that the bonding is partially ionic, without attempting to quantify it.

Where mixed ionic and metallic bonding is present, many such phases crystallize in the fluorite (CaF_2) structure. This structure is one in which the metal (electronegative) ions form a FCC lattice in which the other ions (electropositive) occupy sites which lie upon the volume diagonals at positions one-quarter of the diagonal distance from each corner ion (tetrahedral interstices).

The compounds Mg_2Si, Mg_2Ge, Mg_2Sn, and Mg_2Pb furnish a good illustration of this type of mixed bonding. It is considered that the degree of metallic bonding increases from Mg_2Ge to Mg_2Pb because of the corresponding decreases in the heats of formation of these compounds. This has been ascribed to decreasing electrochemical differences between the components. The heat of formation of Mg_2Si is considerably less than that of Mg_2Ge, being approximately equal to that of Mg_2Sn. This leads to the idea that the bonding of Mg_2Ge may be more ionic in character than that of Mg_2Si. However, the electronegativity differences for the two compounds indicate the reverse.

It has been proposed that as the electronegativity differences of compounds of a series of this kind decrease, the sizes of the ions increase and the wave functions of the valence electrons spread out. This could lead to increased metallic bonding as the valence electrons become associated with increasing numbers of ions.

Some AB phases appear to have mixed ionic, metallic, and covalent bonding. Many compounds of this class crystallize in the NiAs structure. Here the metal ions lie in a HCP sublattice with up to three metalloid ions at c/4 and at 3c/4 (octahedral interstices) in the unit cell. This structure is stable over relatively wide ranges in composition; this factor contributes to the nonhomogeneity of bonding. The A component is usually a transition element and B element is a member of the B subgroups of columns II to VI in the Periodic Table.

The ionic portion of the bonding diminishes as the occupancy of the d levels of the transition element increases. The electronegativity differences also diminish with increasing numbers of d electrons. The smaller electronegativity differences give rise to a smaller degree of ionic bonding. Metallic bonding appears to predominante in many of these compounds because they demonstrate metallic conductivity.

The phases with increasing degrees of ionic bonding have larger axial ratios and narrower ranges of composition with heats of formation of about −8 to −10 kcal/mol. As the degree of ionic bonding decreases, the covalent bonding increases. Under these conditions, the compositional range of the phase decreases. As the degree of covalent bonding increases, the heats of formation of these phases lie in a range of about −9 to −6 kcal/mol. The bonding is primarily metallic in those phases with the smaller heats of formation. As the degree of metallic bonding increases, the compositional ranges of these phases also increase; the metallic constituent can exceed 50%. Some examples of phases with mixed bonding which assume this structure (NiAs) are given in Table 10-15.

Molecular crystals are bound by van der Waals, or London, forces. These arise from instantaneous variation in the symmetry of the charge on a molecule or on an atom; an electric dipole is formed. This induces an electric field in the adjacent molecules or atoms and results in the formation of a second set of dipoles. The interaction between the oppositely charged dipoles gives rise to an attractive force, the van der Waals force. As would be expected, these forces are weak. This is verified by the low melting and boiling points of organic crystals. Crystals of the noble gases also are bound by this means. Bond strengths of this type are estimated to be between -10^{-14} and -10^{-12} erg. These vary as r^{-6} where r is the interdipole distance.

10.6.9. Defect Structures

A defect structure is one, as mentioned in Section 10.6.7, in which some of the lattice sites are not occupied. In lattices such as NiAl, which is BCC, some of the Ni sites remain unoccupied when that element is reduced slightly below 50%.

This can be explained on the basis of the Hume-Rothery scheme of valences (Table 10-14). This assigns a valence of zero to the transition elements. Small variations in the Ni content, therefore, do not change the electron:atom ratio significantly; it remains at $\sim 3/2$. This is the ratio noted previously for the stable β phase. The vacancies remain unoccupied by Al rather than increasing the electron:atom ratio. Similar behavior has been observed in other compounds of transition elements such as CoAl, NiGa, and CoGa. This also has been noted in alloys which do not contain transition elements (Al-base alloys and some γ brasses).

Similar defects are shown by phases of the NiAs type, with strong ionic bonding (Section 10.6.8), which also tend to show vacant sites. These are caused by the absences of the metallic components when their concentrations are reduced. This reduces the degree of metallic bonding and the ionic behavior increases. Decreasing amounts of the metallic component can occur until the composition is given by AB_2. At this composition the lattice type changes.

Table 10-16
ORIGINAL BASIS FOR THE E.-B. RELATIONSHIP

Atom	Ground state	Excited bonding state	Number of bonding (s + p) electrons/ atom	Lattice type
Na	3s	3s	1	BCC
Mg	$3s^2$	3s3p	2	HCP
Al	$3s^23p$	$3s3p^2$	3	FCC

10.7. ENGLE-BREWER RELATIONSHIPS

The Engle-Brewer (E.-B.) work endeavors to correlate the crystal structures of metallic elements and their alloys to the electron configurations of the elements involved. It attempts to extend the Pauling molecular orbital concepts to metals and alloy systems; other factors, such as those discussed earlier in this chapter, are given little, if any, consideration. The Hume-Rothery rules for phase prediction are considered by them to constitute a special case of their model. This approach also attempts to provide a means to correlate the structural behavior of alloys with the electron:atom ratio. It is considered to represent the first stage of an important, but as yet incomplete, empirical contribution to the understanding of alloy phases; many questions still remain unanswered by this approach.

10.7.1. Nontransition Elements

The E.-B. approach considers elements which form structures according to the (8-N) rule as electron-concentration phases. In such cases, the lattice type is determined by the covalent bonding of the valence electrons.

Most metals normally crystallize in one or more of the BCC, HCP, or FCC lattice types. The E.-B. model postulates that these lattices result from the number of bonding electrons. Paired electrons are not considered to be suitable for bonding. The unpaired (s + p) electrons are considered to be suitable for bonding. These unpaired (s + p) electrons alone are considered to affect the lattice type. This is shown in Table 10-16.

Sodium has an s electron available for bonding in the ground state. No excitation, or promotion, of the electronic states is required to acquire an unpaired electron since one already is present. The ground state of Mg is $3s^2$. This outer, paired couplet is not considered to be capable of bonding and must be changed by promoting one of the s levels to a higher, unpaired p state, leaving both outer electrons unpaired so that both may bond. The state requiring the least promotional energy satisfying this requirement is 3s3p.

Energy factors must be taken into account in selecting the excited state. A configuration, such as that given for Mg, is determined by weighing the excitation and bonding energies. An increase in the excitation, or promotion, energy should result in more electrons available for bonding. The absolute value of the bonding energy is assumed to increase as more electrons are made available. The resultant "increase" in bonding energy must more than compensate for the necessary increase in the excitation energy and lower the total energy of the system. This is the reasoning behind the excited bonding states for Mg and Al given in the table. It should be noted that electrons in p half-bands are unpaired.

The BCC, HCP, and FCC lattice types originally were considered to result from 1, 2, and 3 unpaired bonding electrons per atom, respectively. Thus, the alkali metals of column I of the Periodic Table should have BCC lattices; the elements of column II would be expected to form HCP lattices, while those of column III should have FCC lattices. The disparity between the predicted and observed behavior is greatest in column III. It is apparent that additional discriminators were required. The original observations may be summarized as follows:

1. The number of bonds made by an atom is based upon the number of unpaired s and p electrons.
2. Excited electronic configurations must exist in preference to those of the ground state when the bonding energy is "greater" than the excitation energy and lowers the energy of the system.
3. The excited, unpaired (s + p) states are responsible for the lattice type.

10.7.2. Binary Alloy Systems

The above observations considered that the hybridized electron states from the various levels make separable, integral contributions to the bonding. It has long been considered that small valence differences from whole numbers can and do exist (Sections 3.11.2, Volume I and Sections 7.12.1 and 9.9, Volume II). Nonintegral electron configurations often represent the lowest energy cases. This can arise from the overlap of bands and the consequent hybridization of the outer electrons. Evidence for nonintegral valences of normal metals is given by Hall coefficient measurements (Table 11-3). Arguments based upon precise, separate knowledge of the numbers of d and s and/or p states materially weaken the E.-B. case.

Both integral and nonintegral electron configurations have been given for many of the elements. It appears that nonintegral values, in all probability, best represent the configurations of the transition elements. In addition, the excitation energies of some elements crystallizing in the FCC lattice are not known. Here, estimates based upon the integral sp^2 configuration must be used.

The stability, or compositional extent, of the lattice types of various phases were determined on such criteria from selected phase diagrams. These were chosen for monovalent BCC metals paired with divalent HCP metals. The diagrams of divalent metals with trivalent metals also were studied. An effort was made to minimize size effects. This led Engel to the following empirical relationships (Table 10-17).

Thus, it is apparent that the BCC structure is considered to exist with an excess of 0.75 electrons per atom beyond that postulated for the pure element. The FCC lattice can exist with a deficiency of 0.75 electrons per atom compared to the pure metal. The stability range of about ±0.2 electrons per atom for the HCP is the narrowest of the three. It should be noted that the ranges given here, in terms of the electronom ratio, can represent wide composition ranges, depending upon the elements involved.

10.7.3. Transition Elements

The d states of the transition elements are considered to be unpaired up to the point at which the d half-bands are filled at five electrons. When the second half-band begins to fill, these additional electrons pair with those already in the first half-band. The E.-B. approach postulates that the paired d electrons are unavailable for bonding. The unpaired d electrons enter into the bonding process, but are considered *not* to influence the crystal type. The resulting structure, as in the case of the normal metallic elements, is determined *only* by the number of (s + p) electrons.

It is generally agreed that the outer electron levels of the transition elements are hybridized as a result of the overlap of the outer s and d bands because of the facility with which this explains many physical phenomena. This poses a problem because of the different roles postulated for s and d electrons by the E.-B. approach; it is difficult to see how these states can be separated. If it is assumed that such electron states can

Table 10-17
ENGEL RELATIONSHIPS

Lattice type	Elemental electron-atom ratio	Integral electron configuration	Alloy[15] electron:atom ratio
BCC	1	s	<1 to 1.75
HCP	2	sp	1.85 to 2.2
FCC	3	sp^2	2.25 to 3 +

Table 10-18
BONDING ENERGIES (KCAL/ MOL) PER ELECTRON

Shell number	s + p	d
1	80	—
2	40	—
3	20	—
4	16	26
5	15	30
6	15	36

From Engle, N., *Acta Met.*, 15, 558, 1967. With permission. Also see Reference 19.

be separated, another point arises. The E.-B. idea that the d levels contribute significantly to the bonding without affecting the lattice type appears to be inconsistent and has not been sufficiently explained. It has been suggested by Hume-Rothery that the d levels may contribute a van der Waals type of bonding. This must, necessarily be small because of the typically large values of r_a/r_d (Section 9.5, Volume II). The role of the d states is important because of the emphasis placed upon the strength of the d bonds (Table 10-18).

The d bonds are given as being about twice the strength of an outer (s + p) bond, and are comparable to those of single, covalent bonds. Such high-strength bonds are considered to "disrupt" the electron configurations of those transition elements with less than five d states. This is presumed to result in the occupation of d levels by outer states and to cause less than two outer electrons to remain in such states.

As an example, Ti, whose ground state is d^2s^2, was first considered as having a d^3s^2 configuration. E.-B. now approximate this as $d^{2.5}sp^{0.5}$. Apparently, on the average, one-half of one of the s electrons is considered to be promoted to the p band and the other half to the d levels. This configuration is considered to have a bonding energy which more than offsets the excitation energy. The BCC allotrope of Ti requires an excitation energy of about half that needed to excite electrons to the d^2sp array postulated for the HCP lattice. However, their calculations show that the bond strength of the HCP lattice is about 1% greater than that of the BCC. It is, thus, predicted that the low-temperature allotrope should be HCP which transforms to BCC at higher temperatures, despite the fact that the e/a ratio is less than the minimum required for the HCP.

Chromium (ground state: $3d^54s$) is considered to bond without requiring excitation since its d electrons are all in one half-band and unpaired. It, thus, has six bonding electrons only one of which, the 4s, determines the lattice type as being BCC.

Table 10-19
E.-B. ELECTRON CONFIGURATIONS FOR FERROMAGNETIC TRANSITION ELEMENTS

Atom	Ground state	Engle's configuration	Revised configuration	Lattice type	Bohr magnetons
Fe	$3s^23p^63d^64s^2$	$3s^23p^33d^{10}4s$	$3s^23p^63d^74s$	BCC	2.2
Co	$3s^23p^63d^74s^2$	$3s^23p^33d^{10}4s4p$	$3s^23p^63d^74s4p$	HCP	1.7
Ni	$3s^23p^63d^84s^2$	$3s^23p^33d^{10}4s4p^2$	$3s^23p^63d^74s4p^2$	FCC	0.6

Note: See also Table 9-3 (Volume II).

Transition elements with more than five d electrons start pairing at the d levels. The formation of such pairs reduces the number of possible d bonds; this approaches zero as the number of d electrons approaches 10. The E.-B. scheme considers that elements with nearly filled d bands can only increase their number of d bonds if this pairing is decreased. This is accomplished by the promotion of d electrons to outer levels. Such excited, former d states, thus, add d bonds by increasing the number of unpaired d states as well as by contributing outer s or p states to those already present. This gives an (s + p) electron:atom ratio of about 2, or greater.

Transition elements with seven or more d electrons are explained by the excitation of d or s electrons to nominal sp^2 configurations. These are expected to form FCC lattices.

The electron configurations of the elements were estimated from data obtained by spectrographic means. Heats of vaporization were used to approximate bond strengths. Data obtained in these ways were employed to calculate the energies required to raise atoms of postulated electron configurations and known lattice types to the ground state in the vapor phase. For example, where an element has two allotropes the structure with the greatest energy of sublimation is considered to be the more probable. Such a structure has the highest bond strength and, therefore, is considered to be the more stable of the two.

These and other observations were revised later to consider that the BCC lattice would occur where a configuration up to $sp^{0.5}$ is present. The HCP structures are associated with $sp^{0.7-1.1}$. FCC lattices result from $sp^{1.5-2}$ configurations.[19]

10.7.4. Magnetic Effects

Engle considered that Fe, Co, and Ni with BCC, HCP, and FCC lattices should have (s + p) values of 1, 2, and 3 respectively, with similar inner cores. The same core configuration was used to explain the similarity of atomic sizes, melting points, elastic moduli, and ferromagnetism. The bases for the ferromagnetic behaviors are given in Table 10-19.

Engle's configuration is based upon a highly improbable, half-filled p band. Filled p states are extremely stable and it is most unlikely that $3p^3$ states would be present in transition elements. In addition, paramagnetism and ferromagnetism are associated with unpaired d, not p, states (Section 8.6.2 and Chapter 9, Volume II). The revised configurations do attempt to associate ferromagnetism with the d levels. However, none of these electron configurations can account for the observed differences in the magnetic properties of these elements without affecting the prediction of the lattice type. It is apparent that this presents a crucial dilemma.

10.7.5. Allotropy

The E.-B. correlation postulates that the allotropic form with the highest energy of

Table 10-20
THE HUME-ROTHERY EXTRAPOLATIONS

	Element		
Property	Al	Mg	Na
Boiling point	2441°C	1103°C	−235°C (extrapolated) 883°C (actual)
Melting point	660°C	650°C	640°C (extrapolated) 98°C (actual)

sublimation is the most probable since it represents the most stable configuration. The prediction of the BCC lattice is complicated by an entropy effect. Since this crystal type is not close-packed, relatively large vibrational amplitudes can occur. As the temperature increases, the entropy increases, causing a decrease in the Gibbs free energy. Where lattice types appear to show approximately equal bond strengths, the lowering of the free energy can impart a higher probability for the presence of the BCC rather than the HCP or FCC lattices.

The following guides are given for cases where the bonding energy of the BCC is close to that of a more closely packed lattice:

1. When the net bonding energy of the BCC is greater than that of the close-packed structure, the BCC will be the stable lattice.
2. When the net bonding energy of the BCC is smaller than that of the close-packed structure by more than 1 kcal/mol, the entropy factor cannot lower the free energy in favor of the BCC. The close-packed lattice is then predicted.
3. When the bonding energy of the BCC is almost the same as that of the close-packed structure, or is not smaller by more than 1 kcal/mol, the close-packed structure will be probable at low temperatures and the BCC will be probable at higher temperatures. (This results from the entropy factor.)

Iron is considered to be an exception to these rules. The "magnetic contributions to the entropy at lower temperatures are greater than differences in vibrational contributions" associated with the BCC. However, the magnetic transformation in 99.9% Fe takes place over the range from 755 to 791°C. The α-to-γ transformation takes place at 912°C. The higher lattice and magnetic entropies should be expected, on this basis, to stabilize the BCC lattice at temperature below 791°C and the transformation to FCC should take place near this temperature, rather than at 912°C, if the E.-B. ideas are correct. The γ - δ transformation is ascribed to the entropy factor.

10.7.6. Noble Metals

Copper, silver, and gold, whose d bands are filled in the ground state, are treated as transition elements in order to account for their bond strengths and FCC lattices. A large portion of the bonding is attributed to their d states. These are considered to have $d^{9.3}sp^{0.7}$ electron configurations with 2.4 bonding electrons per atom. This conclusion is drawn by means of extrapolations from adjacent transition elements to the noble elements. Hume-Rothery has indicated the dangers inherent in such a procedure by using the elements which formed the original basis of the E.-B. concept. This is shown in Table 10-20. The large disparities between the actual and extrapolated values cast doubts upon such a means for the approximation of bond strengths.

It was first considered that 1.5 d electrons were promoted to provide these elements with an $sp^{1.5}$ configuration. This satisfied the requirement for the FCC lattice and provided both more (s + p) and unpaired d electrons for bonding. This later was modified to a d^8sp^2 configuration. Here, the d states are considered to be paired to avoid the prediction of paramagnetic effects. This configuration would, however lead to results which are not in agreement with the observed properties of these metals and their alloys (see Table 6-2, Sections 7.9.2 and 7.9.4, and Figure 9-17, Volume II). The final $d^{9.3}s^{0.7}$ also is at variance with the observed properties.

These objections were answered by considering that the alloys of the noble metals with the elements in the columns immediately following them in the Periodic Table, which have completed d shells, disturb the d-level behavior postulated for the noble metals. This disturbance is presumed to occur because these alloying elements cannot form d bonds. In this case, the noble metals are considered to return to the monovalent behavior; their alloys with this class of alloying element then follow the Hume-Rothery rules. This is contradictory, since the E.-B. concept predicts BCC lattices for such cases and FCC lattices are observed.

If the d^8sp^2, or any other hybridized electron configuration of noble metals, is considered to be present when they are alloyed with transition elements, further complications arise. Such configurations neither can explain the magnetic properties of alloys of this type, nor their variations as functions of composition. The magnetic variation of the Cu-Ni system as a function of Cu content provides a good illustration of this. It is difficult to consider the noble elements as having transition-like electronic configurations when they are alloyed with transition metals, except as an expedient. The postulation of the d^8sp^2 or $d^{9.3}sp^{0.7}$ configurations for the noble metals apparently is made in order to systemize the relatively large number of isomorphous systems which they form with the transition elements.

10.7.7. Multicomponent Constitution Diagrams

The E.-B. approach, here, assumes that the electron configurations of the atoms involved in the alloy play the predominant roles in alloy phase formation. Plots containing this information are combined with empirical data taken from constitution diagrams and are given in diagrams such as in Figure 10-19.

These diagrams represent the phases over all temperatures up to solidus temperatures. The areas represent the maximum compositional ranges over which the phases exist. They do not account for the temperatures at which the maxima occur. The abscissas denote the atomic composition of the metals shown on the right-hand side. The ordinates are given in terms of *total* numbers of (s + d) electrons.

A horizontal line drawn between the base metal, on the left-hand side, and a second metal on the right, represents the binary diagram of the two metals. In the above illustration, Ir has a total of 9 (s + d) electrons. A line drawn across the diagram linking Ir and W intercepts the BCC area at about 15%. This gives the maximum solid solubility of Ir in W; it does not provide the temperature coordinate. The other phases are treated similarly. A horizontal line drawn across the diagram half way between Os and Ir represents ternary alloys of W with equal atomic percentages of Os and Ir. Various other ratios also can be selected, depending upon the location selected for the horizontal line. It will be noted that Figure 10-19 may be regarded as a condensation of three ternary diagrams of the indicated transition metals.

This approach is based upon the hybridization of the outer electrons. Its application has been most successful for alloys of the transition elements of columns IV, V, and VI with elements belonging to columns VII and VIII of the second and third transition periods.

In more complex systems than those given in the figure, different lattice types may

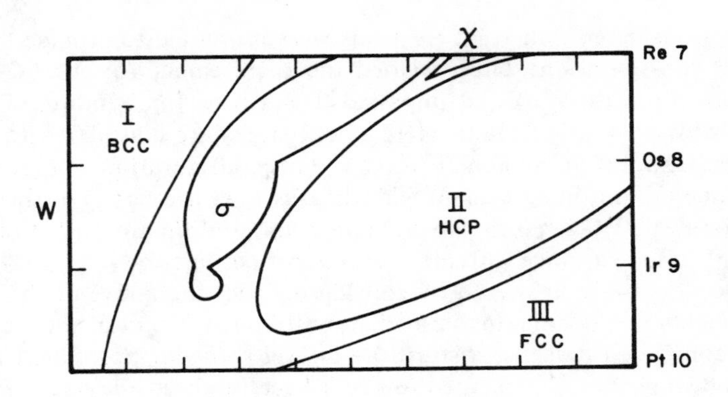

FIGURE 10-19. Multiple phase relationships. (After Brewer, L., in *High Strength Materials,* Zakay, V. F., Ed., John Wiley & Sons, New York, 1965, 92. With permission.)

exist for a given composition, depending upon the temperature. Here, dotted lines are used to outline the area representing the low-temperature phase when the two fields overlap. Such diagrams summarize much information. They can be very helpful, provided that it is remembered that they only are intended to represent the maximum compositional ranges of phase stability and do not provide the corresponding temperatures.

It will be recalled that the existence of a given lattice type was postulated to fall within limiting values of the electron:atom ratio. Some of the boundaries in diagrams, such as Figure 10-19, approximate constant electron:atom ratios. Other boundaries do not correspond to the ratios postulated for the various lattice types. From this, some of the ratios have a rational basis, while others appear to be empirical.

10.7.8. Other Unresolved Questions

The BCC lattices of Li and Na are taken into account by the E.-B. rules. However, these elements have HCP lattices at low temperatures. The close-packed structures are not explained.

Beryllium has a BCC allotrope above 1250°C. The HCP form can be explained by the rules. Be has a ground-state configuration of $1s^2 2s^2$. It is difficult to see how this can be hybridized to provide the 1.5 electron per atom ratio postulated by the rules for the BCC lattice.

The lattice of Ca is FCC. One electron from the p levels must be promoted to its outer levels in order to conform to the rules. However, these completed p levels are very stable and such promotion does not appear to be possible without the application of abnormally large energies. According to the rules, barium, with the same outer configuration, should crystallize as HCP; it crystallizes in the BCC form. Strontium, which also has the same outer configuration as the two previously noted elements, has three allotropes with increasing temperature: FCC, HCP, and BCC. The BCC form, at high temperatures, is explained on the basis of a $d^{0.5}sp^{0.5}$ configuration and lattice entropy. The HCP allotrope is explained by an sp configuration and smaller entropy at lower temperatures. As in the case for the lattice of Ca, the FCC structure has yet to be explained.

Manganese has four allotropic forms. Its ground-state electron configuration is d^5s^2. The excited d^6s and d^4sp^2 states can account for the BCC and FCC allotropes. However, these require that the very stable ground state be changed. A d^5sp structure might be predicted, but Mn has no HCP allotrope. It has been suggested that low bonding energy and the similarity of the energies of excitation for the d^6s and d^5sp configurations might account for the two complex lattices of Mn at lower temperatures.

In summary, the E.-B. concept appears to be the most useful in the prediction of the structures of transition elements and their alloys. The condensed diagrams for the phases present in such systems, while considerably empirical, do provide assistance in the prediction of alloy phases. The E.-B. concept represents the largely empirical beginnings of a contribution to the understanding of alloy phases which still is incomplete. It is possible that this method of correlation might provide a basis for additional, more encompassing, work upon this problem.

10.8. OTHER METHODS OF PHASE DETERMINATION

The Kaufman-Bernstein method for the determination of phase equilibria is a thermodynamic approach and is outlined here. It is based upon the Gibbs free energies of the phases involved and is based upon the ideas of Van Laar.[25]

The general approach to the problem takes into account a comparison of the free energies of all possible phases. Those with the lowest free energies at a given temperature are the most probable and will exist, or coexist, at that temperature. The application of this method is shown schematically, for a binary system in Figure 10-20.

The principal contributions of these investigators consist of the collection and analyses of the pertinent thermodynamic parameters and the use of computer programs to derive the necessary free-energy curves as functions of temperature and composition. These effectively include the Hume-Rothery size factors and electron:atom ratios. Data from the Engle-Brewer correlation are not included because the methods they used to obtain their thermodynamic data are insufficiently well described. Intermetallic phases are treated as having an exact stoichiometry, rather than existing over ranges of composition, as many do (Figure 10-15), in order to make the calculations less complicated. The Pauling electronegativity data are not used for interaction parameters because they give inconsistent results. The degree of success of this method is indicated in Figures 10-21 and 10-22.

It is considered that optimum results might be obtained from a combination of the thermodynamics and solid-state physics.

What appears to be a better approach was suggested by Lomer.[22] His outline of the calculations to be made is: "1, Assume structure; 2, Evaluate self-consistent fields for the component-free atoms; 3, Superimpose these potentials to get band structure, and use the resulting wave function to get self-consistent solution for solid; 4, Evaluate correlation energy correction; 5, Evaluate total energy; and 6, Repeat for all suspected rival structures."

This outline has virtually never been fully applied. The usual procedure is to assume that one of the energy terms is a function of structure, such as E_F. Such an abbreviated approach can account for lattice distortion, or its stability when such criteria as the Hume-Rothery rules are applied; it cannot yield reliable comparisons for the selection of the most probable phase. The reason for this is that omission of the sixth operation is always omitted; assumptions regarding phase stability are not valid substitutes for the required comparisons of energies.

10.8.1. The Miedema Method

At present the most reliable approach to the theory of alloy phases is that of A. R. Miedema.[26] It is a semi-empirical approach based upon the Pauling concept of electronegativity. Equation 10-1 may be expressed as

$$\Delta H \propto -(\Delta x)^2 (eV)$$

An important concept upon which this work is based is that the electronegativity of

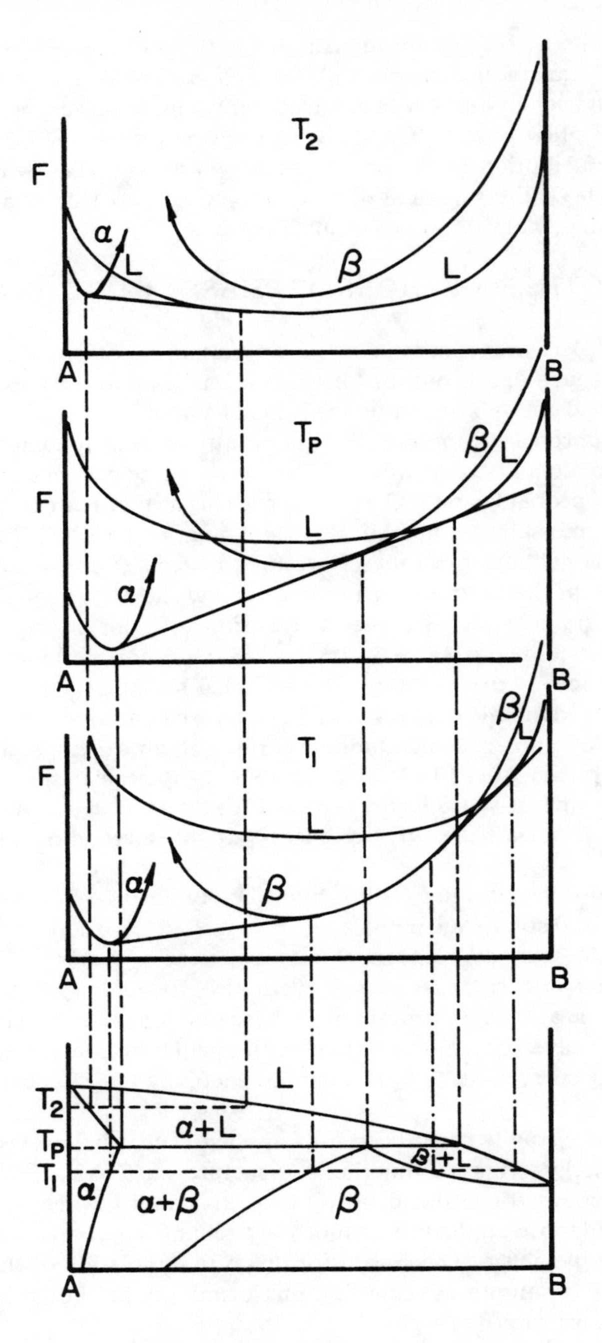

FIGURE 10-20. The determination of a peritectic phase diagram from the free energies of the phases present.

an element is proportional to its chemical potential. (The chemical potential also is directly related to E_F; see Section 7.13, Volume II). Thus, upon substitution,

$$\Delta H \propto -(\Delta\mu)^2 (eV) \tag{10-43}$$

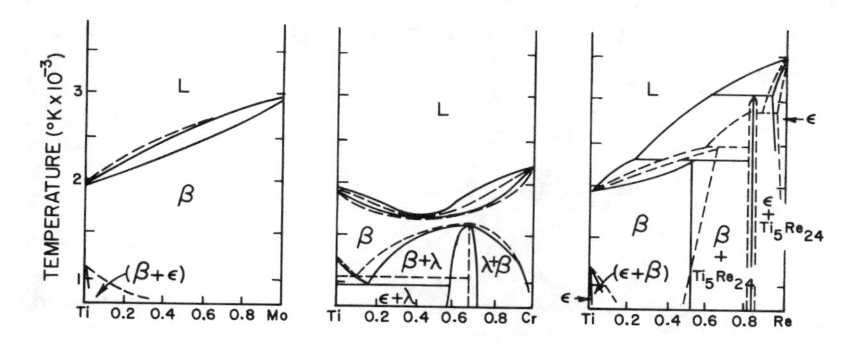

FIGURE 10-21. Comparison of calculated (broken curves) and observed (solid curves) phase diagrams. (After Kaufman, L. and Bernstein, H., *Computer Calculation of Phase Diagrams,* Academic Press, New York, 1970, 193. With permission.)

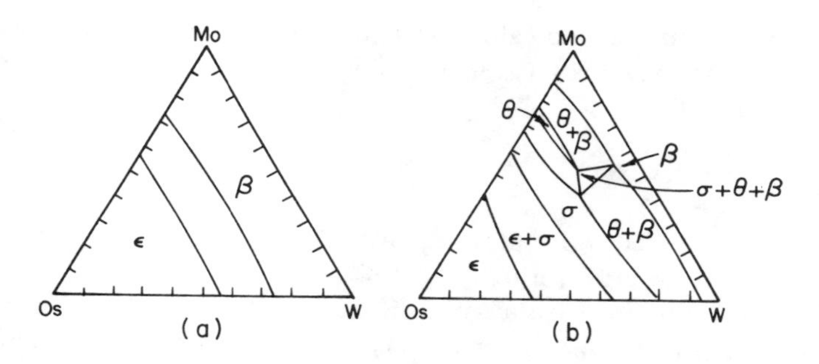

FIGURE 10-22. (a) Calculated and (b) observed ternary diagrams of Mo-W-Os at 2648 K. (After Kaufman, L. and Bernstein, H., *Computer Calculation of Phase Diagrams,* Academic Press, New York, 1970, 240. With permission.)

The chemical potential of a metal, μ_A, is approximately equal to its work function, W_A (see Figure 1-2). (W_A also is related to E_F, section 5.2, Volume I). Thus,

$$\Delta\mu \simeq \Delta W \tag{10-44}$$

and

$$\Delta H \propto -(\Delta W)^2 \simeq -(\Delta\mu)^2 (eV) \tag{10-45}$$

Here ΔW is the difference between the work functions of two metals forming an alloy system.

This means that valence electrons must transfer their association from atoms with smaller work functions to those with larger work functions; changes in E_F, thus, result (Section 7.9.2, Volume II). The resulting change in the electron density is designated as Δn_{AB}. This change affects the "sizes" of the two atoms. The volumes of the Wigner-Seitz proximity cells (Sections 5.8.2, Volume I) also must change because of this. The volume of the cell of an atom which originally was associated with more valence electrons will decrease; that of the other atom which became associated with more electrons will increase. The constitutes another model for the expression of uniform electron density between the two atoms in the lattice of the alloy. The energies of the atoms undergoing such changes are lowered by this mechanism. Thus, $-\Delta\mu \propto \Delta n_{AB}$ and ΔH will be larger:

$$\Delta H \propto (\Delta n_{AB})^2 \tag{10-46}$$

The combination of these two effects gives

$$\Delta H = G(C)\,[-P'(\Delta\mu)^2 + Q(\Delta n_{AB})^2]\ (eV) \tag{10-47}$$

Here G(C) may be a function of composition and P′ and Q are treated as constants, where $P' = Pe$. Equation 10-47 is a thermodynamic relationship. It will be recognized that Equation 10-47 also is a function of E_F because both μ_A and n_A are related to E_F. The values of μ_A, approximated from W_A, show an approximately linear relationship as a function of the Pauling electronegativity (Table 10-3); they fall within a band of about ±0.5 eV on either side of the line of best fit. Thus, data for $\Delta\mu$ are readily available.

Miedema also showed that the value for n_A of a normal metal can be approximated with reasonable accuracy. This gives n_A as a linear function of the compressibility, K, and the molar volume, V_m:

$$n_A \simeq \beta + (KV_m)^{-1/2} \tag{10-48}$$

where n_A is in units of electrons/(atomic unit)^3; 1 a.u. = 0.572 Å. The constant β is slightly larger than $10(KV_m)^{-1/2}$ for $n_A = 10^{-3}$ e/(a.u.)^3.

When $\Delta\mu$, which is obtained from ΔW in Equation 10-44, is plotted as a function of Δn_{AB}, where n_{AB} is expressed in electron density units (1 d.u. $\simeq 6 \times 10^{22}$ e/cm^3), a straight line of demarcation is obtained. This line has a slope of $\Delta\mu/\Delta n_{AB} = 0.48$ and provides a reliable guide to the type of binary alloy system which will be formed between transition elements. This slope gives the relationship Q/P = 0.255 (eV/d.u.)^2. Points representing binary alloy systems which lie to the left of this line have negative values of ΔH and, consequently, form compounds or other stable phases. These phases consist of at least one solid solution with an extent of more than 10 At.% at either low or elevated temperatures. Alloy combinations which lie to the right of the line have positive values of ΔH and do not form stable compounds.

Transition elements alloyed with Cu, Ag, and Au have values of Q/P = 0.23 (eV/d.u.)^2. Alloy systems of transition elements with Li, Ca, and Sr have values of Q/P = 0.27 (eV/d.u.)^2.

In summary, when $\Delta\mu > \Delta n_{AB}$ (to the left of the demarcation line) either compounds of solid solutions will be present. At least one of the solid solutions will have a range of more than 10 At.%. When $\Delta\mu < \Delta n_{AB}$ (to the right of the demarcation line) neither intermetallic phases not ordered phases will be formed. These rules have a reliability of about 98%.

Alloy systems of transition elements and nontransition elements with valences of 3 or greater require an additional term in the equation for ΔH:

$$\Delta H = G(C)\,[-P'(\Delta\mu)^2 + Q(\Delta n_{AB})^2 - R] \tag{10-49}$$

This results in a parabolic rather than a linear line of demarcation between the two types of alloying behavior. In this case Q/P = 0.175 (eV/d.u.)^2. This is significantly different from the ratios noted above. R/P is given as 1.36 eV^2.

It appears that this changed behavior may result from p electrons, since the added elements have 3 to 5 valence electrons. Consideration has been given to the idea that p-d hybridization causes this. However, R is approximately a constant and is not a function of the transition element forming systems with elements contributing p valence electrons. The value of R is the equivalent of a heat of solution of about 1 eV/at, for dilute solutions.

10.9. PROBLEMS

1. Show that the diameter of the largest sphere which will fit into the interstices of an array of close-packed spheres is 0.59 that of a close-packed sphere.
2. Discuss the reasons for the expected changes in the Fermi level and in the zone shape as the composition of a binary, substitutional solid solution varies.
3. Discuss the relative merits of using the distances of closest approach and the Goldschmidt radii in applying the size factor.
4. Verify the predictions of Table 10-2 by reference to constitution diagrams.
5. Compute the electron:atom ratio of the α solid solution of Si in Cu with maximum solid solubility.
6. Make a table similar to Table 10-7 for 5 additional elements forming isomorphous systems.
7. Plot the liquidus curves of the copper-base alloys of Figure 10-8 as functions of their electron:atom ratios.
8. Make a list of additional constitution diagrams which show minima in their liquidus curves and are intermediate between isomorphous and eutectic solidification.
9. Give three examples of the (8-N) rule for each period of the Periodic Table for columns IV to VII, inclusive.
10. Describe and contrast the metallic and covalent bond types. Describe how the metallic bond can be considered to be an extremely dilute covalent bond.
11. Use the values of Table 10-10 to approximate those of Table 10-9.
12. Approximate the bonding energies of the first five compounds of Table 10-11 by means of the approximate form given by Equation 10-19d.
13. Contrast electron phases with ionic phases. Explain their differences as electrical conductors.
14. Compare covalently and ionically bonded phases.
15. Show how some primarily ionic compounds can have a covalent bonding component.
16. Show that the ZnS structure is favored when $r_2 \leqslant 4.55r_1$.
17. Explain in detail why the stoichiometry of Laves phases is important.
18. Give the reasons why the Jones model for electron phases may be considered to be a one-dimensional approach.
19. Use trigonal hybridization to explain the paramagnetic properties of graphite.
20. Explain the properties of diamond in terms of tetrahedral hybridization.
21. Why should III-V compounds form lattices similar to that of the diamond cubic?
22. Discuss the reasons why van der Waals bonds must originate from instantaneous variations in the charge symmetry of molecules or atoms.
23. Show that the electron configurations given in Table 10-19 cannot account for the paramagnetic and ferromagnetic properties of nickel and its alloys.
24. Show that the $d^{9.3}sp^{0.7}$ electron configuration for copper cannot account for its electrical resistivity and those of its alloys.
25. Show that the E.-B. electron configurations cannot account for the thermoelectric properties of Cu-Ni alloys.
26. Critically discuss the Kaufman-Bernstein technique including the major, inherent difficulties.

10.10. REFERENCES

1. Seitz, F., *The Physics of Metals,* McGraw-Hill, New York, 1943.
2. Seitz, F., *The Modern Theory of Solids,* McGraw-Hill, New York, 1940.
3. Hansen, M., *Constitution of Binary Alloys,* McGraw-Hill, New York, 1958.
4. Pearson, W. B., *A Handbook of Lattice Spacings and Structures of Metals and Alloys,* Pergamon, Press, Elmsford, N.Y., 1958.
5. Laves, F., in *Theory of Alloy Phases,* American Society of Metals, Metals Park, Ohio, 1956.
6. Hume-Rothery, W., *The Structure of Metals and Alloys,* Chemical Publishing, New York, 1939.
7. Pauling, L., *The Chemical Bond,* Cornell University Press, Ithaca, N.Y., 1967.
8. Gordy, W., *Phys. Rev.,* 69, 604, 1946.
9. Kittel, C., *Introduction to Solid State Physics,* 3rd ed., John Wiley & Sons, New York, 1966.
10. Mott, N. F. and Gurney, R. W., *Electronic Processes in Ionic Crystals,* 2nd ed., Dover, New York, 1964.
11. Dekker, A. J., *Solid State Physics,* Prentice-Hall, Englewood Cliffs, N.J., 1959.
12. Jones, H., *Proc. R. Soc. (London),* A144, 225, 1934; A147, 396, 1934; *Philos. Mag.,* 43, 105, 1952.
13. Robinson, P. M. and Bever, M. B., in *Intermetallic Compounds,* Westbook, J. H., Ed., John Wiley & Sons, New York, 1967, 38.
14. Barrett, C. and Massalski, T. B., *Structure of Metals,* 3rd ed., McGraw-Hill, New York, 1967.
15. Engle, N., *Acta Met.,* 15, 557, 1967; 15, 565, 1967.
16. Brewer, L., *Acta Met.,* 15, 553, 1967.
17. Hume-Rothery, W., *Acta Met.,* 13, 1039, 1965; 15, 567, 1967.
18. Hume-Rothery, W., *Progr. Mater. Sci.,* 13(5), 231, 1967.
19. Brewer, L., *Science,* 161, 115, 1968.
20. Brewer, L., in *High Strength Materials,* Zakay, V. F., Ed., John Wiley & Sons, New York, 1965.
21. Kaufman, L. and Bernstein, H., *Computer Calculation of Phase Diagrams,* Academic Press, New York, 1970.
22. Rudman, P. S., et al., Eds., *Phase Stability in Metals and Alloys,* McGraw-Hill, New York, 1967, 125.
23. Sanderson, R. T., *J. Chem. Phys.,* 23, 2467, 1955.
24. Cullity, B. D., *Elements of X-ray Diffraction,* Addison-Wesley, Reading, Mass., 1956, 86.
25. Van Laar, J. J., *Z. Phys. Chem.,* 63, 216, 1908; 64, 257, 1908.
26. Miedema, A. R., *Phillips Tech. Rev.,* 33(6), 149, 1973.
27. Pauling, L., *The Nature of the Chemical Bond,* 3rd ed., Cornell University Press, Ithaca, N. Y., 1960, 93.
28. Gordy, W. and Thomas, W. J. P., *J. Chem. Phys.,* 24, 439, 1956.
29. Slater, J. C., *Phys. Rev.,* 23, 488, 1942.

Chapter 11

SEMICONDUCTORS

Semiconductors are good examples of the theoretical and practical applications of modern electron theory. Their electrical resistivities vary in the range from about 10^{-2} to 10^{9} Ω-cm. This range is intermediate between those of good conductors (10^{-6} to 10^{-5} Ω-cm) and insulators (10^{14} to 10^{22} Ω-cm). The elements and compounds discussed here are insulators at 0 K. These materials are covalently bonded. The elemental semiconductors bond according to the (8-N) rule. The compound semiconductors mainly consist of ions from groups III and V and II and VI, but other covalently bonded compounds also have desirable properties (Section 10.6.7). These materials are intrinsic semiconductors in the "pure" state. They have negative temperature coefficients of electrical resistivity.

The modifications of the electrical properties of these materials by means of dopants (extrinsic semiconductors), barriers, surface effects, etc. have led to the development of many types of devices which have had wide application and have revolutionized electronic circuitry.

Some of the more common devices are discussed. Electronic circuits and the techniques employed in the fabrication of integrated circuitry are considered to be beyond the scope of this presentation. However, the reader is provided with a sufficient background to pursue these topics elsewhere.

11.1. INTRINSIC SEMICONDUCTORS

As noted previously, C, Ge, Si, and Sn(α) are covalently bonded and crystallize according to the (8-N) rule. Thus, all of their valence electrons are tightly bound in the bond-sharing process between nearest neighbors. This sharing provides each of the ions with completed outer shells. In other words, the valence bands of these elements are completely filled at 0 K (see Sections 5.7 and 6.7, Volume I, and Figure 5-20, Volume II). The next outermost band, the conduction band, is completely empty of electrons. These two bands are separated by a relatively small energy gap of the order of k_BT. As the temperature is raised above the absolute zero, increasing numbers of electrons are promoted, across the gap, into the formerly empty conduction band. Holes are created simultaneously in the states they vacate in the valence band. As increasing numbers of electrons occupy states in the conduction band, the material becomes increasingly conductive because the many available states in the conduction band permit these electrons to behave in a nearly free way. The holes in the vacated states in the valence band also serve as carriers. Since this process increases with increasing temperatures, the electrical resistivity decreases; these materials have negative temperature coefficients of electrical resistivity. Simply restated, increasing the temperature promotes more carriers from the valence to the conduction band and creates more holes in the valence band. The greater the number of mobile carriers of electricity, the greater the conductivity and the smaller the resistivity of the semiconductor. Thus, the temperature coefficient of resistance is negative.

11.1.1. Intrinsic Conduction Mechanism

The promotion of electrons to the conduction band creates holes in the valence band which also contribute to the conduction process. The vacated states in the valence band, known as holes, are imaginary particles which have a positive charge equal in magnitude to that of an electron. In the case of intrinsic semiconductors, the number of holes, n_h, is equal to the number of electrons, n_e, which is promoted to the conduction band. This is called pair production.

The electrical conductivity, σ, for the case of electrons in a metallic conductor was given by Equation 5-63 (Volume I). This also can be expressed simply as

$$\sigma = ne\mu \tag{11-1}$$

where μ, the mobility is given, as derived from Equation 5-61 (Volume I), by

$$\mu = e\tau/m \tag{11-1a}$$

in which m is the effective mass (Section 11.1.2).

In order to account for the contributions of both electrons and holes to the conductivity of a semiconductor, the effects of both types of carriers must be considered:

$$\sigma = \sigma_e + \sigma_h \tag{11-2}$$

Or, using Equation 11-1,

$$\sigma = n_e e\mu_e + n_h e\mu_h \tag{11-3}$$

It is necessary to obtain expressions for n_e and n_h in order to calculate σ and its temperature dependence. These may be obtained by determining the number of states for each carrier (the area under the curve) such as shown in Figure 5-5c (Volume I). Thus, an expression must be obtained for the density of electron states in the conduction band, $N(E)_c$, and for the density of holes in the valence band, $N(E)_h$. This is done with the help of Figure 11-1.

The density of states in the conduction band is given by

$$N(E)_c = \int_{E_c}^{\infty} N(E)\, f(E)\, dE \tag{11-4}$$

This is the integrated product of Equations 5-21 and 5-50 (see Figure 5-5c, Volume I). The width of the energy span is large with respect to k_BT, so it may be approximated, actually using the Boltzmann tail (Equation 5-50a, Volume I) that

$$f(E) \simeq e^{-(E-E_F)/k_BT} \tag{11-5}$$

When spin is included, the factor N(E) in Equation 11-4 becomes

$$N(E) = \frac{4\pi}{h^3}\,(2m_e^*)^{3/2}\,(E - E_c)^{1/2}\,dE \tag{11-6}$$

Here, m^*_e is the effective mass of an electron at the bottom of the conduction band. The concept of effective mass was introduced in Section 5.8.2 (Volume I) and is discussed in more detail in the next section. The factor V is equal to unity when Equation 5-21 is used here; it does not appear explicitly in Equation 11-6 because n_e and n_h are the numbers of carriers per unit volume. Equations 11-5 and 11-6 are substituted into Equation 11-4 to give the density of states in the conduction band as

$$N(E)_c = \frac{4\pi}{h^3}\,(2m_e^*)^{3/2} \int_{E_c}^{\infty} e^{-(E-E_F)/k_BT}\,(E - E_c)^{1/2}\,dE \tag{11-7}$$

The upper limit of ∞ may be used because only a negligible number of states are filled at energies much above E_c as a result of the very rapid decrease in f(E).

When $E_c - E_F$ is small, the integral can be expressed in the form, since $E_c \sim E_F$,

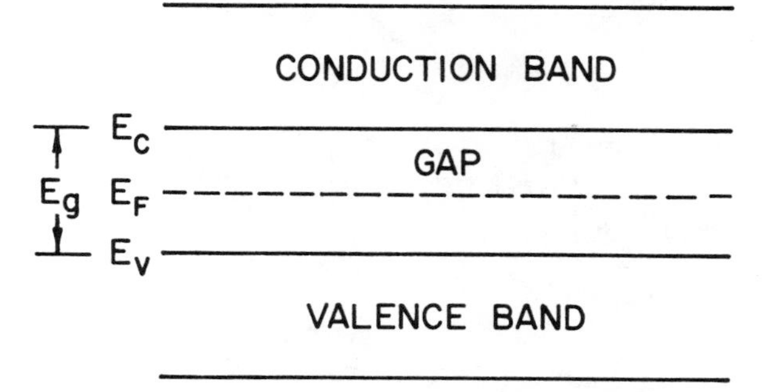

FIGURE 11-1. Simple band model of an intrinsic semiconductor.

$$\int_0^{\infty} e^{-u} u^{1/2}\, du = \frac{\pi^{1/2}}{2}$$

Upon integration it is found that

$$N(E)_c = \frac{4\pi}{h^3}(2m_e^*)^{3/2}\int_{E_c}^{\infty} e^{-(E-E_F)/k_BT}(k_BT)^{1/2}\frac{(E-E_c)^{1/2}}{(k_BT)^{1/2}}\, k_BT\, \frac{dE}{k_BT} = \frac{4\pi}{h^3}(2m_e^*k_BT)^{3/2}\frac{\pi^{1/2}}{2}$$

and the total number of electrons per unit volume in the conduction band is given by, when the expression for $N(E)_c$ is simplified,

$$n_e = N(E)_c f(E) = 2\left[\frac{2\pi m_e^* k_BT}{h^2}\right]^{3/2} \cdot e^{(E_F-E_c)/k_BT} \qquad (11\text{-}8)$$

Since intrinsic behavior is being considered, the probability of finding a hole, or unoccupied state in the valence band is determined by the number of electrons which have entered the conduction band. This is given by [1−f(E)]. The derivation of the density of holes (number of holes per unit volume) in the valence band parallels that given above for the electrons in the conduction band. Thus,

$$N(E)_h = \int_0^{E_v} N(E) \cdot [1 - f(E)]\, dE \qquad (11\text{-}4a)$$

In this case the unoccupied states are close to E_v and [1 − f(E)] quickly becomes negligible as the interior of the valence band is approached. Thus, the density of states for the holes in the valence band is the "mirror image" of that for the electrons in the conduction band and

$$1 - f(E) \simeq e^{-(E_F-E)/k_BT} \qquad (11\text{-}5a)$$

Following the same reasoning as given above

$$N(E)_h = \frac{4\pi}{h^3}(2m_h^*)^{3/2}\int_0^{E_v} e^{-(E_F-E)/k_BT}(E_v-E)^{1/2}\, dE \qquad (11\text{-}6a)$$

in which m^*_h is the effective mass of a hole. Performing the integration, and following the previous treatment, the number of holes per unit volume in the valence band is

$$n_h = N(E)_h[1 - f(E)] = 2\left[\frac{2\pi m_h^* k_B T}{h^2}\right]^{3/2} e^{(E_v - E_F)/k_B T} \qquad (11\text{-}9)$$

The quantities n_e and n_h are in the range of 10^{18} to $10^{19}/m^3$ in the neighborhood of room temperature. These densities are very small compared to the density of electrons, about $10^{25}/m^3$, in a normal metal.

Equations 11-8 and 11-9 can be equated, since $n_e = n_h$ in intrinsic semiconductors. Noting that the coefficients vanish, this gives

$$\exp[(E_F - E_c)/k_B T]\, m_e^{*3/2} = \exp[(E_v - E_F)/k_B T]\, m_h^{*3/2} \qquad (11\text{-}10)$$

It should be noted at this point that m^*_e and m^*_h have been assumed to be single-valued, scalar quantities. These actually are tensor values, known as the density-of-states effective masses (Section 11.1.2). However, this assumption is reasonable for the one-dimensional analysis given here.

When the logarithms of both sides of Equation 11-10 are taken, and the fractions are cleared,

$$E_F - E_c + 3/2\, k_B T \ln m_e^* = E_v - E_F + 3/2\, k_B T \ln m_h^*$$

This reduces to

$$2E_F = E_v + E_c + 3/2\, k_B T \ln m_h^*/m_e^* \qquad (11\text{-}11)$$

Another way of arriving at the relationships between the various energies involved is to consider the widths of the valence and conduction bands to be small with respect to the energy gap between them. The number of electrons per unit volume in the conduction band may be approximated closely by using the Boltzmann tail of Equation 5-50a. Thus,

$$\frac{n_e}{N(E)_c} = f;\ n_e \simeq \frac{N(E)_c}{\exp[(E_c - E_F)/k_B T]} \qquad (5\text{-}50b)$$

since $(E_c - E_F)/k_B T \gg 1$. And, in the same way, the density of holes in the valence band is

$$n_h \simeq \frac{N(E)_v}{\exp[(E_v - E_F)/k_B T]} \qquad (5\text{-}50c)$$

In both of these equations the total number of states per unit volume is N. From these equations

$$n_e + n_h \simeq \frac{N(E)_c}{\exp[(E_c - E_F)/k_B T]} + \frac{N(E)_h}{\exp[(E_v - E_F)/k_B T]} = 2N$$

Simplifying, and using the series expansion for e^x, gives

$$\frac{1}{(E_c - E_F)/k_B T + 1} + \frac{1}{(E_v - E_F)/k_B T + 1} = 2$$

This may be simplified further to read

$$\left[\frac{E_c - E_F + k_BT}{k_BT} + \frac{E_v - E_F + k_BT}{k_BT}\right]^{-1} = 2$$

since $k_BT \ll E_c - E_F$ or $E_v - E_F$, and finally,

$$E_F = \frac{E_v + E_c}{2} = \frac{E_g}{2} \qquad (11\text{-}12)$$

Therefore, the Fermi level lies at the center of the energy gap between the valence and conduction bands (Figure 11-1).

The substitution of Equation 11-12 into Equation 11-11 gives

$$E_F = \frac{E_g}{2} + \frac{3}{4} k_BT \ln m_h^*/m_e^* \qquad (11\text{-}13)$$

When the bands are exactly symmetrical, $m^*_h = m^*_e$ and Equation 11-13 gives the result

$$E_F = \frac{E_g}{2} \qquad (11\text{-}14)$$

which is the same as Equation 11-12.

It should be noted that m^*_h generally is larger than m^*_e (see Table 11-2). However, even when m^*_h/m^*_e is significantly different from unity, the temperature dependence of E_F is small because Equation 11-13 contains the logarithm of this ratio. Thus, in general, E_F has a small, positive slope. Silicon has $m^*_h/m^*_e < 1$ and therefore has a small, negative slope and is an exception to the more general behavior.

The identical Equations 11-12 and 11-14 enable the simplification of Equation 11-8 as

$$n_e = 2\left[\frac{2\pi m_e^* k_BT}{h^2}\right]^{3/2} \exp(-E_g/2k_BT) \qquad (11\text{-}15a)$$

and Equation 11-9 as

$$n_h = 2\left[\frac{2\pi m_h^* k_BT}{h^2}\right]^{3/2} \exp(-E_g/2k_BT) \qquad (11\text{-}15b)$$

The information now is available for the use and more complete understanding of Equation 11-3. Equations 11-15a and 11-15b may be substituted into Equation 11-3 to obtain

$$\sigma = 2\left[\frac{2\pi m_e^* k_BT}{h^2}\right]^{3/2} \exp(-E_g/2k_BT)\, e_e\mu_e -$$

$$2\left[\frac{2\pi m_h^* k_BT}{h^2}\right]^{3/2} \exp(-E_g/2k_BT)\, e_h\mu_h \qquad (11\text{-}16)$$

The minus sign appears because the sign of the charge on the hole is opposite to that on the electron. Since these have the same absolute value, Equation 11-16 can be expressed as

$$\sigma = 2\left[\frac{2\pi m^* k_B T}{h^2}\right]^{3/2} \cdot |e| \cdot \exp(-E_g/2k_BT)(\mu_e - \mu_h) \qquad (11\text{-}17)$$

if it is assumed that $m_e^* \simeq m_h^*$.

When the first three factors of Equation 11-17 are included in a parameter, A, since $T^{3/2}$ is small compared to the exponential term, this equation may be expressed in logarithmic form as

$$\ln\sigma = \ln A - \frac{E_g}{2k_BT} + \ln(\mu_e - \mu_h) \qquad (11\text{-}18a)$$

or as

$$\ln\sigma = \ln A - \frac{E_F}{k_BT} + \ln(\mu_e - \mu_h) \qquad (11\text{-}18b)$$

If the temperature dependencies of the mobilities of the electrons and holes are neglected, and the mobilities are included with the other factors in the parameter, A, then

$$\ln\sigma = \ln\sigma_o - \frac{E_F}{k_BT} \qquad (11\text{-}19)$$

A schematic plot of this equation is given in Figure 11-2.

The solid curves in the figure show the type of temperature dependence predicted earlier for the intrinsic behavior of semiconductors. The broken curves show the corresponding properties for extrinsic materials which are discussed later in Section 11.2.

The temperature dependence of the mobilities of the carriers was neglected in Equation 11-19. A complete presentation of carrier mobility is beyond the scope of this text. A simple description is given here. The electrons involved are at very low levels in the conduction band. This means that their wave vectors are very small. Consequently, their wavelengths are large and their frequencies and velocities are small. The phonon scattering in pure, intrinsic semiconductors is approximately proportional to $T^{3/2}$. The impurity scattering of electrons varies approximately as $T^{-3/2}$ in extrinsic semiconductors. The mobility varies as $T^{-3/2}$ for intrinsic materials and as $T^{3/2}$ for extrinsic materials. The inverse relationship between scattering and mobility may be understood by examining the relationships $\mu = e\tau/m$ and $\tau = L/v$ (Equation 5-64, Volume I). From these $\mu = eL/mv$. L is the mean-free path which results from scattering and is an inverse function of T. Therefore, the mobility is inversely proportional to the scattering. The mobility of holes varies approximately as $T^{-5/2}$. These relationships are also approximately valid for the Hall effect (Table 11-4).

11.1.2. Electron Behavior and Effective Mass

The masses of electrons and holes were taken into consideration in the previous sections and in Section 5.8.2 (Volume I). The effective masses of these particles give a more accurate representation than that of nearly free particle behavior in certain situations. In the case of normal metals, where the Fermi levels are well below the tops of their valence bands, the effective mass is approximately equal to the nearly free electron mass. Thus, the relationships given in Chapters 5, 6, (Volume I) and 7 (Volume II) are essentially correct. However, in the case of semiconductors, where electrons and holes are very close to the top of the valence band, their effective masses must be taken into account. Effective masses are designated by m*, as in previous discussions.

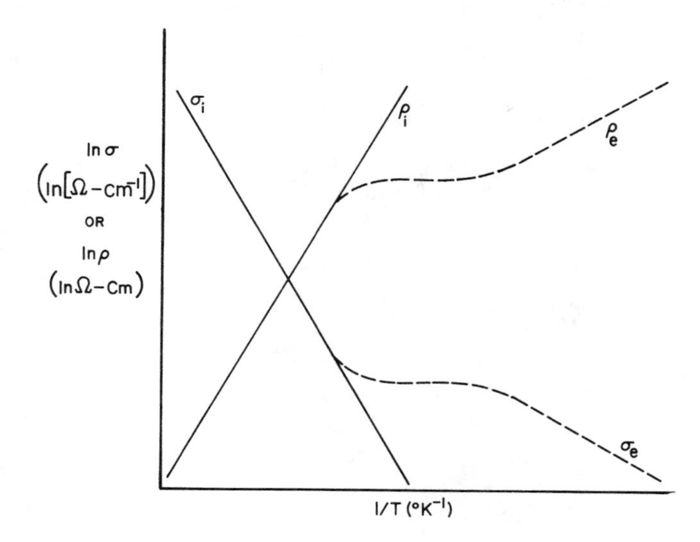

FIGURE 11-2. Electrical conductivity and resistivity of an intrinsic semiconductor (solid curves). Extrinsic behavior is indicated by broken curves.

The behavior of electrons close to the top of the band, very near the zone boundary, will be examined first. The electron is considered as a wave packet. The group velocity is obtained from Equation 2-13 (Volume I) for one dimension, using the wave vector instead of the wave number. Thus,

$$v = v_g = \frac{d\nu}{dk} = \frac{\partial \omega}{\partial \bar{k}_x} \tag{11-20}$$

Recalling Equation 1-6 (Volume I), $E = h;\nu = h\omega/2\pi$, and differentiating with respect to $\bar{k}_x$ result results in

$$\frac{\partial E}{\partial \bar{k}_x} = \frac{h}{2\pi}\frac{\partial \omega}{\partial \bar{k}_x} \tag{11-21}$$

Rearranging Equation 11-21 gives

$$\frac{\partial \omega}{\partial \bar{k}_x} = \frac{2\pi}{h}\frac{\partial E}{\partial \bar{k}_x} \tag{11-21a}$$

The substitution of Equation 11-21a into Equation 11-20 provides a more convenient expression for the velocity of an electron:

$$v = \frac{2\pi}{h}\frac{\partial E}{\partial \bar{k}_x} \tag{11-22}$$

Thus, the velocity of an electron depends upon the slope and consequently, its position on the E vs. $\bar{k}_x$ curve (Figure 5-22b, Volume I). An expression for $\partial E/\partial \bar{k}_x$ is obtained from the derivative of Equation 3-26 (Volume I) as

$$\frac{\partial E}{\partial \bar{k}_x} = \frac{h^2 \bar{k}_x}{4\pi^2 m} \tag{11-23}$$

Now, the velocity of an electron with E and $\bar{k}_x$ values on the parabolic portion of the curve is obtained by the substitution of Equation 11-23 into Equation 11-22.

$$v = \frac{2\pi}{h} \frac{h^2 \bar{k}_x}{4\pi^2 m} = \frac{h\bar{k}_x}{2\pi m} \tag{11-24}$$

Therefore, in the range in which an electron is nearly free, its velocity is a linear function of $\bar{k}_x$ (Figure 11-3a).

However, at and beyond an inflection point on the E vs. $\bar{k}_x$ curve near the top of a band (close to the zone boundary) in Figure 5-22b (Volume I), $\partial E/\partial \bar{k}_x$ decreases and approaches zero at the critical values of $\bar{k}_x$ which occur at $\pm n\pi/a$. The velocity (Equation 11-24) therefore, also approaches zero between the inflection points and $\pm n\pi/a$. This is shown in Figure 11-3b.

The acceleration of an electron, a, is obtained from the derivative of Equation 11-22 with respect to time:

$$a = \frac{dv}{dt} = \frac{2\pi}{h} \frac{\partial}{\partial t} \left[\frac{\partial E}{\partial \bar{k}_x} \right] \tag{11-25}$$

The indicated operation is performed and reexpressed as

$$a = \frac{2\pi}{h} \frac{\partial^2 E}{\partial \bar{k}_x \partial t} = \frac{2\pi}{h} \frac{\partial^2 E}{\partial \bar{k}_x^2} \frac{\partial \bar{k}_x}{\partial t} \tag{11-26}$$

An expression for $\partial \bar{k}_x/\partial t$ is needed. This is obtained by considering the effect of an applied electric field, E, upon an electron. If the field influences the electron for a time dt, and its velocity is v, the distance it travels in that time is vdt. Hence,

$$dE = e\bar{E}vdt \tag{11-27}$$

Equation 11-27 may be reexpressed as

$$\frac{dE}{dt} = \frac{\partial E}{\partial \bar{k}_x} \cdot \frac{\partial \bar{k}_x}{\partial t} = e\bar{E}v \tag{11-27a}$$

An expression for $\partial E/\partial \bar{k}_x$ is obtained from Equation 11-22. This is substituted into Equation 11-27a with the result being

$$e\bar{E}v = \frac{h}{2\pi} v \frac{\partial \bar{k}_x}{\partial t} \tag{11-28}$$

Rearrangement of Equation 11-28 will provide the expression for $\partial \bar{k}_x/\partial t$ being sought:

$$\frac{\partial \bar{k}_x}{\partial t} = \frac{2\pi}{h} e\bar{E} \tag{11-28a}$$

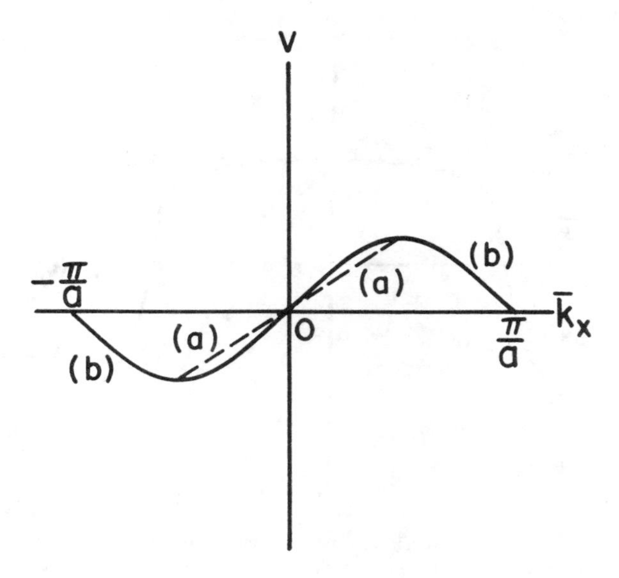

FIGURE 11-3. Velocity as a function of wave vector; (a) or a nearly free electron; (b) the influence of the zone boundary.

The substitution of Equation 11-28a into Equation 11-26 gives the acceleration of an electron as

$$a = \frac{2\pi}{h} \cdot \frac{\partial^2 E}{\partial \bar{k}_x^2} \cdot \frac{2\pi}{h} e\bar{E} = \frac{4\pi^2}{h^2} \cdot \frac{\partial^2 E}{\partial \bar{k}_x^2} \cdot e\bar{E} \qquad (11\text{-}29)$$

A nearly free electron which is not close to the top of the band, or to the zone wall, that is, in the parabolic portion of the E vs. $\bar{k}_x$ curve (as in a normal metal), will have its acceleration determined by the derivative of Equation 11-23. For these conditions

$$a = \frac{4\pi^2}{h^2} \cdot \frac{h^2}{4\pi^2 m} \cdot e\bar{E} = \frac{e\bar{E}}{m} \qquad (11\text{-}30)$$

so that the acceleration of a nearly free electron is constant for a given electric field (Figure 11-4a).

The acceleration of an electron near the top of the band is more complicated; it is more complex since $\partial^2 E/\partial \bar{k}_x^2$ (Equation 11-29) no longer remains constant. This factor is positive in the parabolic portion of the E vs. $\bar{k}_x$ curve, becomes zero at the inflection points and is negative from the inflection points to the zone boundaries. The acceleration, accordingly, decreases from a positive value, goes through zero and then becomes negative (Figure 11-4b).

The effective mass of an electron may be determined from an examination of Equation 11-29:

$$a = \frac{F}{m} = \frac{4\pi^2}{h^2} \cdot \frac{\partial^2 E}{\partial \bar{k}_x^2} \cdot e\bar{E}$$

It can be seen from this equation that the effective mass is given by

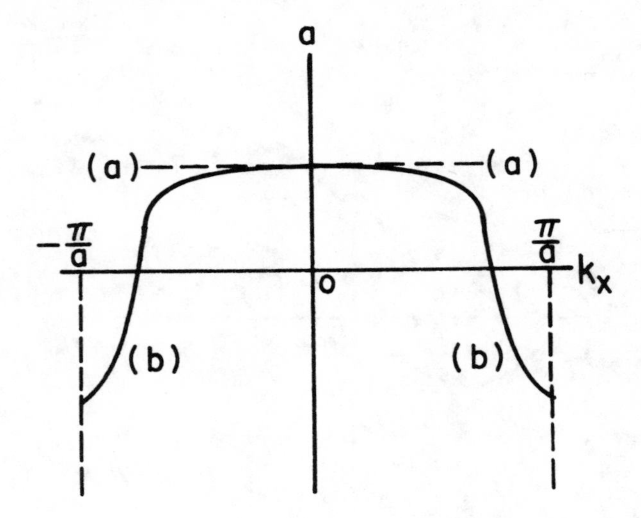

FIGURE 11-4. Acceleration as a function of wave vector: (a) for a nearly free electron; (b) the influence of the zone boundary.

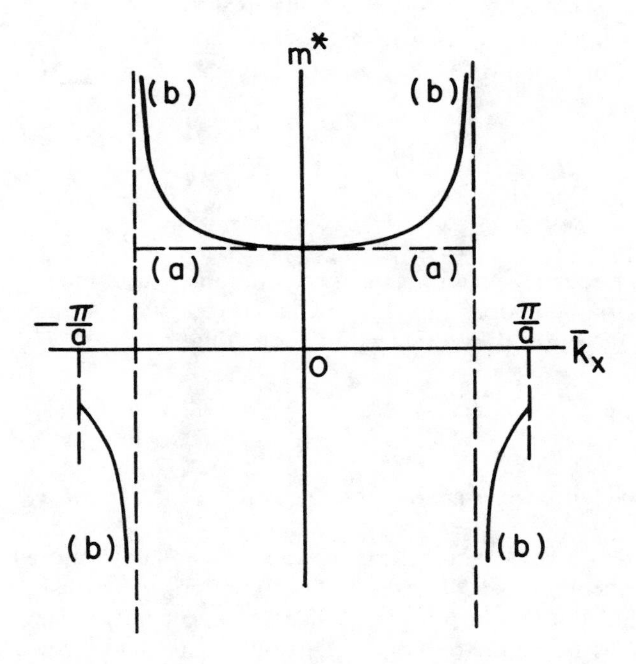

FIGURE 11-5. Effective mass as a function of wave vector: (a) for a nearly free electron; (b) the influence of the zone boundary.

$$m^* = \frac{h^2}{4\pi^2}\left[\frac{\partial^2 E}{\partial \bar{k}_x^{\,2}}\right]^{-1} \tag{11-31}$$

An inspection of Figures 5-22b or 5-23 (Volume I) and Equation 11-23 shows that $\partial^2E/\partial\bar{k}_x^2$ is positive and constant in the parabolic part of the curve and $m^* \simeq m$, the mass of a nearly free electron. This is shown in Figure 11-5a.

At the inflection points of the curve of E vs. $\bar{k}_x$, the second derivative equals zero. This produces singularities. Beyond this, as the zone boundary is approached, m* is negative The singularities actually represent maxima rather than infinite masses. This results from the Heisenberg Uncertainty Principle (Equation 2-35, Volume I). Since the energy cannot be known exactly, average values of m* are determined over a suitable ΔE. These result in maxima or minima.

The curve of E vs. $\bar{k}_x$ for holes in a nearly filled band may be considered to be the mirror image of that for electrons (Figure 11-6) because their energies, charges, and masses are opposite in sign. Thus, $\bar{k}_x$ for a hole equals zero at the top of the valence band where the maximum wave vector of an electron would have been at $\pm\pi/a$. Further, the energies of the holes increase parabolically downward, symmetric about $\bar{k}_x = 0$, as the magnitudes of their wave vectors increase. This gives the sign of $\partial^2E/\partial\bar{k}_x^2$ for a hole as opposite to that for an electron. This is responsible for the effective mass of a hole being opposite in sign to that of an electron on a given portion of the curve.

In the case of insulators, where the filled valence bands and empty conduction bands are separated by large energy gaps, conduction is not possible. Neither the application of normal fields nor temperatures can promote electrons across the gap. Thus, under most conditions, the effective mass of electrons in insulators need not be considered. However, in certain cases (for example, Section 11.8.1) suitable radiant energy can promote valence electrons to the conduction band and these materials behave in the same ways as semiconductors.

The application of external energy to a semiconductor results in the utilization of a part of the applied energy in the promotion of some of the electrons from the valence to the conduction band. The carriers remaining in the valence band also are affected. Consider the case in which an electric field is applied parallel to and in the $+\bar{k}_x$ direction (Figure 11-3b). Let one unfilled state with a $-\bar{k}_x$ coordinate be present close to the top of the valence band. An electron in the next highest state would be expected to occupy the unfilled state; its energy and velocity component parallel to the field should be decreased in the direction opposite to the field (Lenz's law). This electron behaves as expected, fills the hole and leaves the hole in the state it vacated. Thus, the hole effectively moves from a higher state to the next lower state; it moves in the same direction as the applied field, as would be expected of a particle with a positive charge. Its velocity is positive because its value of $\partial E/\partial\bar{k}_x$ is opposite in sign to that of electrons. Its acceleration and effective mass are positive because $\partial^2E/\partial\bar{k}_x^2$ is positive.

The electron which occupied the original hole in the band will have its negative velocity decreased because it obeyed Lenz's law (Figure 11-3b). This results since Equation 11-23 is negative and similarly affects Equation 12-24. It will have a negative effective mass because $\partial^2E/\partial\bar{k}_x^2$ is negative (Equation 11-31 and Figure 11-5b). A subsequent occupation of the newly created hole by an electron from another slightly lower state transports the hole to that slightly lower state. Following the same reasoning as above, the velocity of the electron will be slightly less negative, its acceleration will be slightly less negative, and its effective mass will be slightly less negative. Correspondingly, the velocity of the hole will increase slightly, its acceleration will be greater, and its effective mass will be slightly more positive.

This description is incomplete until the effective mass of an electron is considered in terms of its reflection (umklapp), at the Brillouin zone boundary. When the Fermi level is close to $\bar{k}_c$, the electrons with energies near E_F will be accelerated by an applied electric field. Some of these electrons will have $\bar{k}_x \geqslant \bar{k}_c$ and will be reflected back into the zone (Section 5.8 Volume I). The reversal of motion changes the sign and magnitude of $\bar{k}_x$. The direction of motion is opposite to that which normally would be expected to have been induced by the field. The concept of effective mass is employed because it readily accounts for this unusual behavior of electrons near the top of the

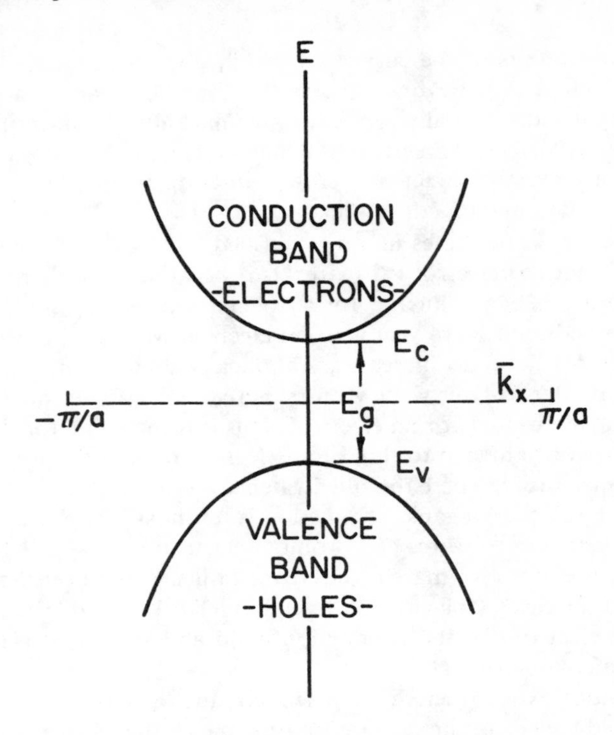

FIGURE 11-6. Curves of energies as functions of wave vector for a small number of holes in a nearly filled valence band and for a small number of electrons at the bottom of an empty conduction band.

band. This simplifies matters by considering that the electron has changed its mass so that its motion remains in conformity with Newton's laws. The change in the mass of the electron is a function of $\bar{k}$. The reversal of the sign of $\bar{k}$ and the corresponding changes in the velocity and acceleration of the electron are properties associated with negative effective mass.

In reality no electrons have negative effective masses. Newton's laws have not been violated, since only the reaction of the electron to the applied field has been considered; the influence of the lattice forces upon the electron have been neglected. The effective mass actually is a crystal-wave interaction; it is not an inertial effect. It is necessary to be concerned with negative effect masses only for those electrons in states near the top of the band, the region in which the E vs. $\bar{k}_x$ curve deviates from its initial, parabolic shape. States with lower energies, where the energies lie upon the parabolic portion of this curve, may be considered to be nearly free and $m^* \simeq m$.

A vacant state, or hole, remains when an electron is excited from the top of a nearly filled band. Another excited electron, from a slightly different state, fills this hole. In so doing, it leaves behind an unoccupied state at its original energy level. After a series of such reactions, the promotion of a series of electrons to higher energy states results in the corresponding displacement of the hole to lower energy states. The particles move in opposite directions. The hole has a positive charge (equal and opposite in sign to that of an electron) and, consequently, is expected to behave in a manner opposite to that of an electron. The hole has a positive effective mass where the electron in the corresponding state has a negative effective mass. Since the holes mirror the behavior of the electrons in a nearly filled band, and in fact control this, it is more convenient to describe the behavior of the carriers by describing that of the holes in the nearly filled valence band. This gives further insight into Figure 11-6 and the discussions which followed.

It will be appreciated that the one-dimensional treatment given here represents the simplest case. Real lattices are anisotropic. In one such case, assuming ellipsoidal energy surfaces, the effective mass may be obtained from Equation 3-49 (Volume I) by letting $L_i = \lambda_i/2$, and recalling that $\bar{k}_i = 2\pi/\lambda_i$, giving

$$E = \frac{h^2 \bar{k}_x^2}{8\pi^2 m_x} + \frac{h^2 \bar{k}_y^2}{8\pi^2 m_y} + \frac{h^2 \bar{k}_z^2}{8\pi^2 m_z}$$

and the effective mass is given by

$$1/m^* = 1/m_x^* + 1/m_y^* + 1/m_z^*$$

This is the "density-of-states" effective mass. In the most general case, the density-of-states effective mass must be given by a nine-component tensor. However, Equation 11-31 can be used where the Fermi energy surfaces may be approximated reasonably well as being spherical. Therefore, m may be used in place of m* in the equations based upon nearly free electrons, given in Chapters 5, (Volume I) and (Volume II) 6, and 7 without introducing appreciable errors.

Normal metals have partially filled valence bands where the occupied states lie upon the parabolic portion of the curve of E vs. $\bar{k}$, with E_F well below the inflection point. Their Fermi surfaces may be approximated as being spherical (Section 10.6.6). This permits the approximation $m^* \simeq m$ and that the electrons are nearly free (see Table 11-1).

Transition elements have narrow, incompletely filled d bands which overlap broad, incompletely filled s bands, and hybridization results. An empty quasi-continuum exists above the highest filled state. This permits the assumption that the valence electrons are nearly free and the approximation that $m^* \simeq m$. This is not as good an approximation as that for normal metals. Even though s-d hybridization is present, the d states are more strongly bound to the nucleus than the s states, so this approximation must be used with caution.

It will be appreciated from the previous discussions that properties involving the mass of the carrier must reflect the nature of the band structure of the crystalline solid. This is particularly true of semiconductors, where both electrons and holes are involved in different bands (Figure 11-7). This gives the variations in the band structure of silicon for two different crystallographic directions. The effective masses of the carriers in some of the more common semiconductor materials are given in Table 11-2.

Bands separated by relatively small energy gaps, as is the case for many semiconductors, have large curvatures in their curves for E vs. $\bar{k}$, (Figure 5-22b, Volume I) adjacent to the zone boundaries. This causes electrons in states at or near the top of the valence band to possess small, negative values for m^*_e. Electrons in such states are represented by the holes by means of m^*_h. States at the bottom of the conduction band have small, positive values of m^*_e as shown in Table 11-2.

11.1.3. The Hall Effect

When a constant magnetic field is applied to a material carrying an electric current, additional momentum is imparted to the carrier; this is normal to the field and to the original momentum of the charged carriers. The trajectories of the carriers become curved and this causes them to concentrate at, or near, the top or bottom surfaces of the specimen, depending upon their charges. The case of an extrinsic semiconductor

Table 11-1
EFFECTIVE MASSES OF ELECTRONS IN SOME NORMAL METALS

Alkali Metals		Noble Metals	
Metal	m*/m	Metal	m*/m
Li	1.19	Cu	0.99
Na	1.0	Ag	1.01
K	0.99	Au	1.01
Rb	0.97		
Cs	0.98		

Adapted from Girifalco, L. A., *Statistical Physics of Materials*, John Wiley & Sons, New York, 1973, 108.

is shown in Figure 11-8, in which the majority carriers are electrons. Here, the electrons become concentrated at the bottom surface of the specimen and leave an excess of positive charge at the opposite surface.

A typical electron transporting the current prior to the application of the field has a velocity of v_e. The magnetic field, H, induces a force on the electron

$$F_m = -ev_e x H \tag{11-32}$$

which tends to make it travel toward the bottom surface of the specimen (the −y direction). No current path exists in the y direction, so the electrons begin to accumulate along the bottom surface and create an excess of positive charge upon the opposite surface. This build-up of charges creates an electric field which exerts a force

$$F_e = e\bar{E}_y \tag{11-33}$$

which tends to oppose that induced by the magnetic field. The total force acting upon the electrons is given by the sum of Equations 11-32 and 11-33 as

$$F = e\bar{E}_y - ev_e x H \tag{11-34}$$

At dynamic equilibrium (steady state), Equation 11-34 equals zero and

$$e\bar{E}_y = ev_e H \tag{11-35}$$

The current density is obtained from Equation 5-4 (Volume I) as

$$j = nev_x$$

From which

$$v_x = \frac{j}{ne} \tag{11-36}$$

Equation 11-36 is substituted into Equation 11-35 to obtain

Table 11-2
PROPERTIES OF SOME LIGHTLY DOPED SEMICONDUCTORS AT 300 K

Type	Semiconductor	E_g (eV)	μ_e (cm²/Vsec)	μ_h (cm²/Vsec)	m^*_e/m_o	m^*_h/m_o
Element IV	C	5.3	1,800	1,600	—	—
	Si	1.1	1,350	475	0.23	0.12
	Ge	0.7	3,900	1,900	0.03	0.08
	SiC	2.8	400	50	0.60	1.20
III-V	AlS	2.2	180	—	—	—
	AlP	3.0	80	—	—	—
	AlSb	1.6	200	420	0.30	0.40
	BN	4.6	—	—	—	—
	BP	6.0	—	300	—	—
	GaAs	1.4	8,500	400	0.07	0.09
	GaP	2.3	110	75	0.12	0.50
	GaSb	0.7	4,000	1,400	0.20	0.39
	InAs	0.4	33,000	460	0.03	0.02
	INP	1.3	4,600	150	0.07	0.69
	INSb	0.2	80,000	750	0.01	0.18
II-VI	CdS	2.6	340	18	0.21	0.80
	CdSe	1.7	600	—	0.13	0.45
	CdTe	1.5	300	65	0.14	0.37
	ZnS	3.6	120	5	0.40	—
	ZnSe	2.7	530	16	0.10	0.60
	ZnTe	2.3	530	900	0.10	0.60
IV-VI	PbS	0.4	600	200	0.25	0.25
	PbSe	0.3	1,400	1,400	0.33	0.34
	PbTe	0.3	6,000	4,000	0.22	0.29
II-IV	Mg_2Ge	0.7	530	110	—	—
	Mg_2Si	0.8	370	65	—	0.46
	Mg_2Sn	0.4	210	150	—	—
II-V	Cd_3As_2	0.1	—	15,000	0.05	—
	CdSb	0.5	300	1,000	0.16	0.10
	Zn_3As_2	0.9	—	10	—	—
	ZnSb	0.5	10	350	0.15	—

Adapted from Wolf, H. F., *Semiconductors,* Interscience, New York, 1971, 33. With permission.

$$e\bar{E}_y = e \frac{j}{ne} H$$

or the Hall field as

$$\bar{E}_H = \frac{1}{ne} jH \tag{11-37}$$

This defines the Hall coefficient as

$$R_H = 1/ne = \mu/ne\mu = \mu\rho \tag{11-38}$$

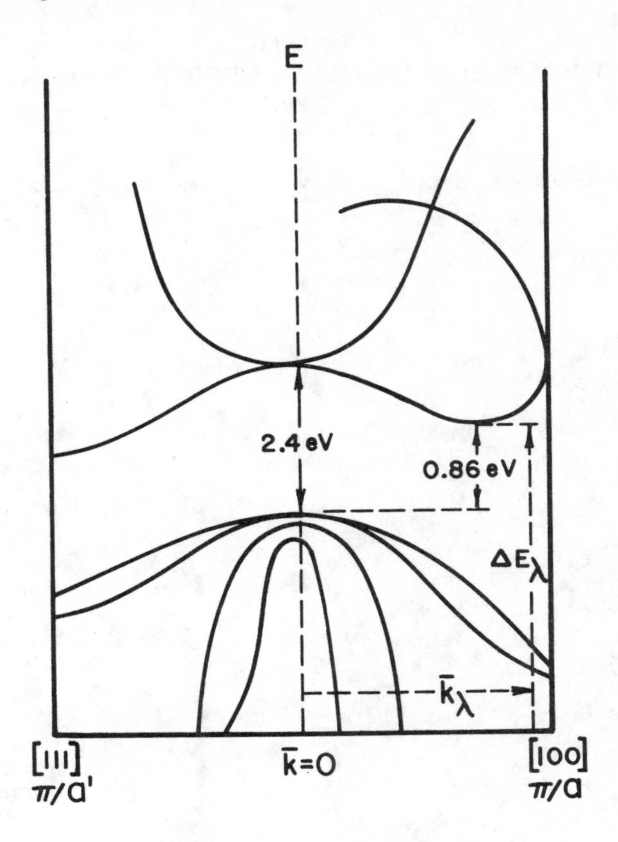

FIGURE 11-7. Approximate band structure of silicon as a function of crystal direction.

In intrinsic and nondegenerate extrinsic materials, $R_H = 3\pi/8ne$ to account for phonon scattering.

The mobility, defined on the basis of Equation 5-61 (Volume I), is given by

$$\mu = v_x/\bar{E}_x; \; v_x = \mu_H \bar{E}_x$$

Here, μ_H is the Hall mobility. This is substituted into Equation 11-35 to give

$$e\bar{E}_y = e\mu_H \bar{E}_x H$$

Upon rearrangement, the Hall mobility is found to be

$$\mu_H = \frac{\bar{E}_y}{\bar{E}_x H} \tag{11-39}$$

The Hall voltage may be obtained from Equation 11-37 as

$$V_H = R_H IH/d \tag{11-40}$$

in which I is the current and d is the thickness of the specimen. The accuracies of determinations of R_H made in this way depend to a large extent upon the probes being

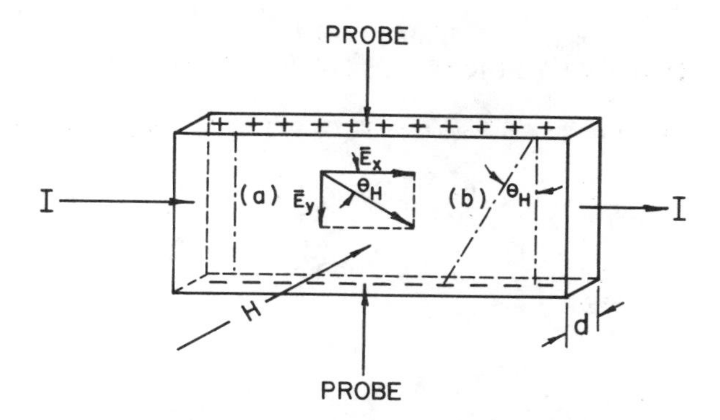

FIGURE 11-8. Sketch of Hall effect. H is normal to the face of the specimen. (a) and (b) represent equipotential surfaces before and after the application of the field, respectively.

as accurately aligned as possible. It also is necessary that the potential across the probes, V_H, be measured using null-balance circuits so that no current flows between the probes. In addition, it is important that the value of d be constant for a given specimen and be known accurately.

The values of n are large for metals and alloys and vary only slightly with temperature. Their values of V_H and R_H, consequently, are small. Such measurements may be made easier by making d as thin as practical. If the current is increased, Joule heating will occur; measurements made under this condition must be made in a constant-temperature bath, or its equivalent, so that extraneous voltages are not included in the measured values of V_H (see Chapter 7, Volume II).

In the case of semiconductors, n is relatively small and is a function of temperature (Equations 11-15a and 11-15b) so V_H and R_H can be very large. Thus, the precautions noted for metals are not quite as important here, except for temperature control. These specimens may be prepared by photochemical etching thin wafers of the material to be tested. The probes and current leads for the measurements of V_x, V_H, and I may be made to be an integral part of the specimen. Wires are spot welded to terminal areas at the ends of the probes and leads in order to connect the specimen to the appropriate measuring devices in the circuit.

Where such specimens are unavailable, μ_H may be calculated from measurements of the Hall angle, θ_H. This is the angle that the equipotential surfaces are rotated by the influence of the magnetic field. Specimen holders can be made so that one of the probes used to measure V_H may be moved along the edge of the specimen until the potential between the probes is zero. The angle between the two probes is θ_H. For small angles $\theta_H = \bar{E}_y/\bar{E}_x$. The substitution of this into Equation 11-39 gives $\mu_H = \theta_H/H$.

The Hall coefficient frequently is given in esu. Where this is done

$$R_H = 1/nec \tag{11-38a}$$

where $c \simeq 3 \times 10^{10}$ cm/sec. Where both electrons and holes are involved in the conduction mechanism, the Hall coefficient is given by

$$R_H = \frac{1}{ec} \cdot \frac{n_h\mu_h^2 - n_e/\mu_e^2}{(n_h\mu_h + n_e\mu_e)^2} \tag{11-38b}$$

The subscripts e and h refer to electrons and holes, respectively. In intrinsic semiconductors, where $n_h = n_e$ and μ_h is usually much smaller than μ_e, Equation 11-38b reduces to the equivalent of Equation 11-38a:

Table 11-3
HALL COEFFICIENTS OF SOME METALS

Element	R_H ($\times 10^{12}$)[a]	Element	R_H ($\times 10^{12}$)	Element	R_H ($\times 10^{12}$)
Cu	−5.5	Be	24.4	Fe	100
Ag	−8.4	Zn	3.3	Co	24
Au	−7.2	Cd	6.0	Ni	−60
Li	−17.0	Al	−3.0		
Na	−25.0				

[a] V/cm-abamp-gauss.

From Seitz, F., *Modern Theory of Solids*, McGraw-Hill, New York, 1940.

$$R_H \simeq -1/nec \tag{11-38c}$$

Measurements of the Hall effect are of both theoretical and practical utility. They provide the numbers and signs of the carriers and their mobilities. If the Hall coefficient is negative, as given either by Equation 11-38a or Equation 11-38c, the principle carriers are electrons; if it is positive, these are holes. This has been used to calculate the effective number of valence electrons per atom in metallic alloys.

It will be noted that Equation 5-63 (Volume I) for electrical conductivity, involves the square of the charge of the carrier and, thus, cannot be used to discriminate between the signs of the carriers. Hall measurements can do this.

The Hall mobilities provide a useful means to determine the scattering of carriers by impurities, imperfections, and phonons. This parallels the discussion given at the end of Section 11.1.1.

Equation 11-38c agrees reasonably well with the model used for normal metals, in which the bands, or zones, are considerably less than completely filled. In the case of the alkali metals, the Hall coefficients are very close to calculations based upon an electron:ion ratio of unity. This is also approximately true for the noble metals, but the agreement is not quite as good as that for the alkali metals. Metals with higher valences show lesser degrees of agreement with the theory; the problem is that the calculation of the number of valence electrons is more complex. Such elements, especially the transition elements, may have nearly filled bands and both hole and electron conduction can be involved.

Antimony and bismuth have large Hall coefficients which are in agreement with Equation 11-38a. These elements have significant diamagnetic properties, (Table 8-1, Volume II); this implies that their bands are nearly filled. Relatively few electrons appear to be available in outer states. The Hall coefficient of Bi is negative; that of Sb is positive.

Hall data are of great use in the manufacture of semiconductor devices. These are used to determine the degrees of lattice perfection, the purity of intrinsic materials, and to verify the actual quantities of alloying elements (dopants) in solution in extrinsic materials. The Hall coefficients of some metals are given in Table 11-3. Those of semiconductors can vary widely, depending upon the kinds and amounts of impurities present. The Hall mobilities of some semiconductors are given in Table 11-4. These data also are affected by the impurity content.

11.2. ROLES OF IMPURITIES (EXTRINSIC SEMICONDUCTORS)

The effects of impurities in solid solution in semiconductors are directly related to the ways in which they enter into the bonding mechanism of the host materials (Section

Table 11-4
HALL MOBILITIES IN SELECTED SEMICONDUCTORS

Semiconductor	μ_e[a]	μ_h[a]
Si	1,300 $(300/T)^{2.0}$	500 $(300/T)^{2.7}$
Ge	4,500 $(300/T)^{1.6}$	3,500 $(300/T)^{2.3}$
GaAs	8,500 $(300/T)^{1.0}$	420 $(300/T)^{2.1}$
GaP	110 $(300/T)^{1.5}$	75 $(300/T)^{1.5}$
GaSb	4,000 $(300/T)^{2.0}$	1,400 $(300/T)^{0.9}$
InAs	33,000 $(300/T)^{1.2}$	460 $(300/T)^{2.3}$
InP	4,600 $(300/T)^{2.0}$	150 $(300/T)^{2.4}$
InSb	78,000 $(300/T)^{1.6}$	750 $(300/T)^{2.1}$

[a] cm^2/V-sec.

Adapted from Wolf, H. F., *Semiconductors,* Interscience, New York, 1971, 303. With permission.

10.6.7) and affect their band structures. Semiconductors containing very small, controlled amounts of intentionally added "impurities", or alloying elements, are known as extrinsic semiconductors. Such additions are called "dopants".

It will be recalled that the elements which crystallize in the diamond-cubic lattice have valences of four; this makes possible the tetrahedral bonding shown in Figure 10-17. An ion from group V of the Periodic Table will have one more valence electron than is required for this type of bonding. The extra electron constitutes a local negative charge in the lattice, near the impurity ion; this can be excited to the conduction band when the temperature is such that k_BT is sufficient to promote it there from a state in the gap, leaving a positive ion behind. Such ions are called donors because they contribute electrons to the conduction process. Materials doped in this way are called n-type semiconductors.

Ions from Group III lack one electron for the required, saturated, covalent bonding. An electron from one of the adjacent semiconductor ions will "orbit" about both the group-III and the ion with which it originally was associated. This induces a positive charge upon the group-III ion. The positive charge acts as a potential well and can capture an electron from a nearby semiconductor ion. Thus, one of the covalently bonded ions loses an electron in this process; it becomes incompletely bonded. The lack of an electron on the trivalent (group III) ion has caused a hole to transfer to the semiconductor ion, leaving the trivalent ion negatively charged. The hole now may be transferred from one semiconductor ion to another in the valence band, without requiring the application of additional energy, because each such state is indistinguishable from the first state. Therefore, the hole can move through the lattice, in the valence band, and contribute to the conduction process. Ions which create such positive holes are known as acceptors. Semiconductors containing such dopants are called p-type because the extrinsic conduction is a result of the positive holes.

Another class of dopants can act either as donors or acceptors. These are known as amphoteric dopants. For example, copper may act as either of these in Ge, depending upon its degree of ionization. Another way in which amphoteric behavior can occur is by the addition of a tetravalent ion to a III-V compound. If it substitutes for a group V ion, it will behave as an acceptor; it will act as a donor if it replaces a group III ion. A third mechanism for amphoteric behavior depends upon the position of the dopant ion in the host lattice. It may act as an acceptor or as a donor depending upon whether it is in a substitutional or interstitial site.

The general effect of the introduction of any of these types of dopants into the semiconductor material is to create new energy states. These levels may lie anywhere between the top of the valence band and the bottom of the conduction band. In any event, the doped semiconductor will be an insulator at 0 K. At higher temperatures, some of the electrons from donor levels can absorb the thermal energy and jump to the conduction band. Such electrons behave as though they were nearly free because many vacant states are available to them. The small number of electrons in the conduction band produces an increase in the electrical conductivity (Figure 11-2) because more carriers become available. This continues up to a limiting condition, the exhaustion range, which is discussed below. Such materials are n-type semiconductors because the principle carriers are electrons. These materials have negative Hall coefficients (Section 11.1.3).

The acceptor dopants behave in a way similar to that of the donors. The capture of an electron from a semiconductor ion will occur only when the thermal energy (k_BT) is sufficient to excite an electron from the valence band to an acceptor state which lies within the gap, but is sufficiently close to the top of the valence band for the exchange to take place. Increasing numbers of holes are created in the valence band as this process continues with increasing temperatures. Since the number of holes increases, so does the electrical conductivity (Figure 11-2). This mechanism also can continue up to a limiting range, the exhaustion range. Materials of this kind are p-type semiconductors because the majority carriers are holes; these materials have positive Hall coefficients. Simple band models of n- and p-type semiconductors are shown in Figure 11-9.

It should be noted that the positive charges on donor ions are not holes and that the negative charges on acceptor ions are not electrons. These are charges on those dopant ions which engaged in giving up or taking on electrons. Therefore, the two conduction mechanisms may be considered as resulting from ionization processes which provide the majority carriers. The ionized dopant ions can move only by diffusion in the lattice of the semiconductor. Thus, they may be considered as not entering into the conduction mechanism.

Both types of impurity levels are localized in the gap and are finite in number. The exhaustion ranges, which were indicated previously, are reached when virtually all of the donors, or acceptors, become ionized. This occurs at relatively low temperatures because the energy levels of the dopant ions are relatively close to the conduction or valence bands. These energy differences are small compared to the energy range of the gap. At temperatures below the exhaustion ranges, the number of intrinsic carriers of either type are in the minority. After exhaustion occurs, the electrical conductivity changes very slowly until a temperature is reached at which the thermal energy is sufficient to promote increasing numbers of electrons from the top of the valence band into states low in the conduction band; increased intrinsic behavior, then, is responsible for the further increase in electrical conductivity (Section 11.1.1). Greater numbers of both electrons and holes participate in the conduction process.

In the case of intrinsic behavior, E_F was approximated as being at $E_g/2$ (Equation 11-12). The reason for this is that the probability of finding a carrier, either an electron or a hole, is 0.5 at that location. However, in the case of extrinsic materials, this is no longer true because of the presence of dopant states within the gap. The Fermi level for either type of dopant can be determined from the number of carriers of a given kind per unit volume in the given band. In the case of donors, this number of electrons is obtained by use of the Boltzmann tail to give, in the same way as Equation 5.50b in Section 11.1.1,

$$n_e \simeq N(E)_c \exp\left[-(E_c - E_{F(d)})/k_BT\right] \qquad (5\text{-}50d)$$

where n_e and $N(E)_c$ are the number of electrons and the density of states in the conduc-

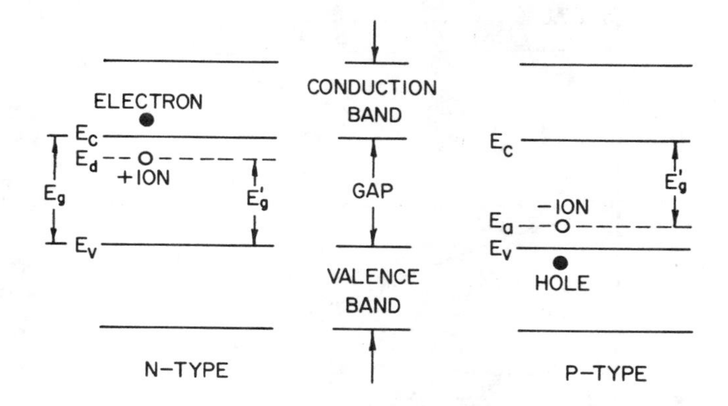

FIGURE 11-9. Simple band models for n- and p-type semiconductors.

tion band, respectively, and $E_{F(d)}$ is the Fermi level in an n-type material. Similarly, following Equation 5-50c in Section 11.1.1,

$$n_h \simeq N(E)_v \exp\left[-(E_v - E_{F(a)})/k_B T\right] \tag{5-50e}$$

in which n_h, $N(E)_v$, and $E_{F(a)}$ have the meanings for p-type material which correspond to those given for Equation 5-50d. Referring back to intrinsic behavior, and using Equation 11-8 to find the density of states in the conduction band, it is found that

$$N(E)_c = n_i \exp\left[-(E_F - E_c)/k_B T\right] \tag{11-41}$$

where n_i denotes the number of intrinsic carriers per unit volume. Equation 11-41 is substituted into Equation 5-50d to give

$$n_e \simeq n_i \exp\left[-(E_F - E_c)/k_B T\right] \exp\left[-(E_c - E_{F(d)})/k_B T\right]$$

This simplifies to

$$n_e \simeq n_i \exp\left[-(E_F + E_{F(d)})/k_B T\right]$$

or

$$E_{F(d)} - E_F = k_B T \ln(n_e/n_i) \tag{11-42}$$

And, in a similar way, but using Equation 11-9, it is found that

$$E_F - E_{F(a)} = k_B T \ln(n_h/n_i) \tag{11-43}$$

Equations 11-42 and 11-43 show the effects of dopants and temperature upon the Fermi level. The effects of dopant concentration, at constant temperature, are shown in Figure 11-10.

The effects of both dopant concentration and temperature are shown in Figure 11-11. The low concentrations of carriers are of the order of $10^{14}/cm^3$ and the high concentrations are about $10^{18}/cm^3$. At low dopant concentrations and at low temperatures $E_{F(d)}$ and $E_{F(a)}$ lie between the impurity states and either the conduction or the valence

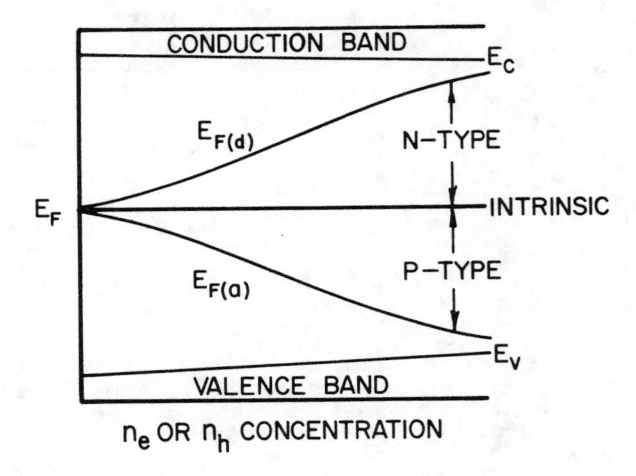

FIGURE 11-10. Variation of the Fermi energy as a function of types of dopant concentration at a constant temperature

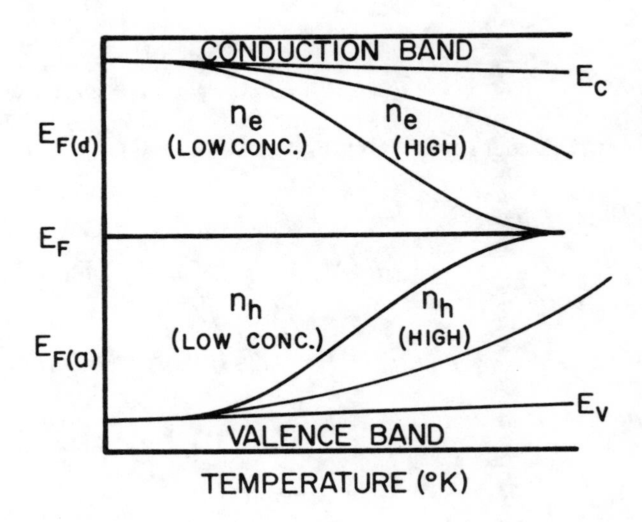

FIGURE 11-11. Variations of the Fermi energy as a function of types of dopant concentration and temperature.

bands. As the temperature increases, increasing amounts of extrinsic behavior take place and both $E_{F(a)}$ and $E_{F(d)}$ begin to approach the intrinsic Fermi level. As exhaustion is approached, this increases until, at that temperature at which intrinsic behavior begins to predominate, $E_{F(d)}$ and $E_{F(a)}$ become identical to E_F. Materials with high dopant concentrations have high exhaustion temperatures and therefore approach this limiting condition at higher temperatures than do materials with low dopant concentrations.

Approximations of the energy levels of the donors and acceptors may be made. These are based upon their ionization energies in solid solution in the covalently bonded semiconductor crystal. Consider a group III or group V ion, with either one less or one more electron than a host ion. The charge difference, either a positive hole or a negative electron, is not tightly bound to the dopant ion, but exists as a "cloud" around it; this is spread over a relatively large number of host ions (say Si or Ge) which surround it. If, for example, a phosphorous ion is in solution in Si, it is assumed that the average distance of the extra electron is relatively far from the P ion. It may be approximated that the radius of the spherical volume around an impurity ion in which electrical neutrality exists is equal to the Debye length:

$$L_D = (\epsilon_D k_B T/2e^2 n_i)^{1/2}$$

where ε_D is the relative dielectric constant. Under this set of conditions, the host lattice may be approximated as being continuous and to possess a uniform dielectric constant. The problem of the calculation of the ionization energy of the extra electron now is similar to the calculation of the bonding energy of an electron in the case of hydrogen. The potential energy of the extra electron is reduced because the surrounding polarized silicon decreases the positive field of the P ion (Section 12.1), hence the large radius; so

$$\text{P.E.} = -e^2/\epsilon_D r \tag{11-44}$$

where r is used as in Section 1.5 (Volume I) instead of L_D. Its total energy is

$$E_d = \text{P.E.} + \text{K.E.} = -e^2/\epsilon_D r + 1/2mv^2 \tag{11-45}$$

An expression for r can be obtained from

$$F = e^2/\epsilon_D r^2 = ma = mv^2/r$$

which reduces to

$$r = e^2/\epsilon_D mv^2 \tag{11-46}$$

An expression for the velocity may be obtained by considering that the electron travels in a closed path about the P ion. This leads, as in Equation 1-30 (Volume I), to

$$\oint P(\theta)\, d\theta = 2\pi p(\theta) = nh = 2\pi mvr \tag{11-47}$$

in which p (θ), the momentum, is constant since there is no torque. From Equations 11-47 and 11-46:

$$nh = 2\pi mvr = 2\pi mv \frac{e^2}{\epsilon_D mv^2} = \frac{2\pi e^2}{\epsilon_D v}$$

And, solving for v,

$$v = \frac{2\pi e^2}{\epsilon_D nh} \tag{11-48}$$

Now, substituting Equation 11-48 into Equation 11-46 gives

$$r = \frac{e^2}{\epsilon m} \cdot \frac{\epsilon_D{}^2 n^2 h^2}{4\pi^2 e^4 \epsilon_D} = \frac{\epsilon_D n^2 h^2}{4\pi^2 e^2 m} \tag{11-49}$$

Equations 11-48 and 11-49 are substituted back into Equation 11-45 to obtain

$$E_d = -\frac{e^2}{\epsilon_D} \cdot \frac{4\pi^2 e^2 m^*}{\epsilon_D n^2 h^2} + \frac{m^*}{2} \cdot \frac{4\pi^2 e^4}{\epsilon_D{}^2 n^2 h^2}$$

This reduces to

$$E_d = -\frac{4\pi^2 e^4 m^*}{\epsilon_D{}^2 n^2 h^2} + \frac{2\pi^2 e^4 m^*}{\epsilon_D{}^2 n^2 h^2} = -\frac{2\pi^2 e^4 m^*}{\epsilon_D{}^2 n^2 h^2} \qquad (11\text{-}50)$$

Here m* has been substituted for m for reasons given in Section 11.1.2. Equation 11-50 is essentially the same relationship obtained from Equation 1-34 (Volume I) for Z = 1 and a relative dielectric constant of unity. It is the binding energy of the extra electron divided by the square of the relative dielectric constant of the medium in which it is traveling. When appropriate substitutions are made (Section 1.5, Volume I), Equation 11-50 becomes, in terms of Equation 1-34a,

$$E_d = -\frac{13.6\ m^*}{\epsilon_D{}^2 m}\ (\text{eV}) \qquad (11\text{-}50a)$$

It is seen that the inclusion of the relative dielectric constant significantly lowers the bonding energy of the extra electron below that given for hydrogen in Section 1.5. It is interesting to note that E_d is independent of the dopant ion. It is a function only of the host material. The "radius" of the extra electron is about two orders of magnitude greater than that of the H atom, so that it most probably is far from the donor ion. Therefore, any local effects introduced by the donor do not affect the electron; it only "sees" the host material. This is the reason why ε_D, a macroscopic quantity, may be used.

If ε_D for Si is taken as being about 12 and $m^*/m \sim 0.5$, then $E_d \simeq 0.047$ eV. This is in good agreement with the observed values for P donor states in Si. These lie about 0.045 eV below the bottom of the conduction band.

Another interesting observation is that despite the fact that Equation 11-50a was derived for a donor ion, it holds for acceptor ions as well. In fact, this equation could have been derived for an acceptor ion instead of a group V ion. This is shown by the fact that the observed value for boron states is $E_a = 0.045$ eV above the top of the valence band in Si. Similar calculations for Ge, using $\varepsilon_D = 16$ and $m^*/m = 0.8$, give $E_d = E_a \simeq 0.043$ eV. The corresponding experimental values for P and B in solution in Ge are $E_d = 0.044$ eV and $E_a = 0.045$ eV, respectively. For the case of a III-V compound doped with a group IV ion, which, as discussed earlier, can show amphoteric behavior, using $\varepsilon_D = 12.5$ and an average $m^*/m \simeq 0.3$, gives $E_d \simeq 0.026$ eV below the conduction band. This is in excellent agreement with the experimental value for Si acceptor levels in GaAs. The relatively small values of E_d and E_a compared to those of E_g (see Table 11-2) explain the onset of extrinsic semiconduction at temperatures so very much lower than that of intrinsic behavior; the close proximity of the impurity levels to the valence and conduction bands requires little thermal energy for the transitions. Levels such as these are designated as shallow levels. Those levels farther from band edges are classed as deep levels.

11.2.1. Generation, Recombination, and Trapping of Carriers

The generation of carriers by thermal means was discussed in Section 11.1 for intrinsic materials and in Section 11.2 for extrinsic materials. These represent the simplest mechanisms. Other, more complicated, processes are possible. Real crystals contain both impurities and imperfections. These may constitute states within the band gap. Thus, the promotion of an electron from the valence band in a nearly intrinsic semiconductor need not necessarily be a single-step process; it also may proceed by one or a series of intermediate steps which involve states lying within the gap. See Figures 11-12a, b, and c.

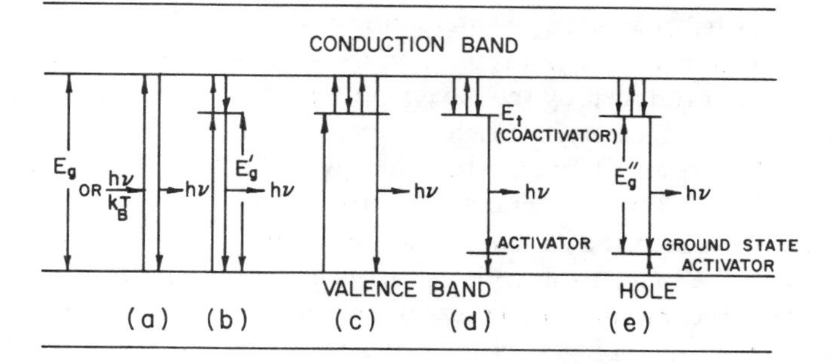

FIGURE 11-12. Recombination processes: (a) direct recombination; (b) indirect recombination; (c) indirect recombination by multiple steps; (d) phosphorescence at a luminescent center; (e) recombination at a luminescent center. Corresponding behavior is shown by acceptor states.

Electromagnetic radiation also may promote an electron to the conduction band. The behavior of real materials is more complex than that shown in Figure 11-12. Figure 11-7 is used to describe this mechanism. The lowest states in the conduction band of Si occur at values of $\overline{k}$ closer to the zone boundary than to its center along the [100] axis. These states correspond to momenta greater than zero ($\overline{k} > 0$). States at the top of the valence band, along the same axis, have zero momentum ($\overline{k} = 0$). An incident photon must have an energy greater than 0.86 eV (the minimum band gap), because it will impart momentum to both the electron and the hole, and must conserve momentum. The photon would require a minimum energy of 2.4 eV for direct generation. The wave vector of incident radiation required for other indirect pair production is indicated by $\overline{k}_\lambda$ in Figure 11-7 (see Section 11.3).

The Auger effect provides another means of excitation of carriers. When two similarly charged particles collide, one of them recombines with a particle of opposite sign and the remaining particle has its energy increased by an amount given up by the one which recombined. This energy transfer can promote the uncombined particle to the appropriate band, creating an electron-hole pair (see Section 11.6.1).

Each of the above mechanisms is reversible. In a thermally activated, intrinsic semiconductor, the electron may drop back to the valence band and recombine with a hole. In so doing, it will give up a photon and/or a phonon. The generation and recombination of electron-hole pairs is continuous and a net equilibrium number of carriers will exist at a given temperature or set of radiation conditions in the intrinsic range. This is important in some of the compound semiconductors.

Multistep generation and recombination may take place by means of the states within the gap. In Si and Ge, an electron may drop to a recombination center within the gap and emit a phonon or a photon. This may be considered to be the equivalent of the emission of a hole. The recombination center may be an ionized impurity (either substitutional or interstitial), or a lattice defect (a dislocation, vacancy, etc.). Recombination may take place at such centers if a hole already is present, or if a hole is captured subsequently. Thermal activation then may promote the electron out of the center and enable it to drop back to recombine with a hole in the valence band, with the release of additional energy. The exit of the electron from the center is the equivalent of the introduction of a hole.

The direct recombination of a particle which initially was excited by a photon will result in the emission of a photon corresponding to its k_λ. This mechanism is important in many compound semiconductors (also see Section 11.8.2).

A particle promoted by an Auger interaction will release the energy it received as a result of the initial collison when it returns to its original state. It should be noted that the number of pairs produced by the Auger process is very small under normal conditions. It is only when large electric fields are applied that this mechanism can create sufficient pairs. This results from the fact that the fields are required to add sufficient energy to the carriers so that their energies approach E_g.

It also should be noted that a carrier captured by a recombination center may be promoted back to the band from which it arrived at the center, rather than return to its original band. Recombination centers acting in this way are known as traps. For example, an electron dropping to a localized trap state may be excited back to the conduction band and then drop to another trap, etc. This process may be repeated several times before the electron drops to the valence band once more (Figure 11-12c). Mechanisms involving traps are indirect recombinations. It is apparent that traps may absorb and emit both electrons and holes. The probability that a trap will capture a carrier depends largely upon its $\bar{k}$ value, and, consequently, upon its energy or its velocity. This is discussed more fully below.

The ability of a recombination center to absorb a carrier depends upon its degree of ionization. This determines the binding energy which is a function of the depth of the potential well which captures the carrier. The lower the degree of ionization, the shallower the well and the smaller the binding energy. For example, if such a state had its energy close to the bottom of the conduction band, an electron could readily leave it by thermal activation. These are shallow traps. The greater the degree of ionization, the deeper the well and the closer the state is to the center of the gap. Therefore, ionized states closer to the center of the gap are more effective recombination centers. These are known as deep traps or recombination centers.

States close to the bands are more efficient in trapping. The probability of an electron leaving a trap is given by the Boltzmann factor: $\exp(-E_t/k_BT)$, where E_t is the depth of the trap. If $n(\nu)$ is a vibrational function associated with the frequency of a classical particle in the well constituted by the trap (Equation 3-56, Volume I) some of these oscillations could free the electron. The rate of escape from the trap is given by $n(\nu)\exp(-E_t/k_BT)$. The function $n(\nu)$ is about 10^8/sec at ordinary temperatures.

Therefore, E_t is a major factor in the determination of the probability of exciting a carrier out of a trap state. The probability that a carrier will be made unavailable by a trap or a recombination center also depends upon the extent of the wave functions possessed by these states. Thus, the probabilities of excitation of a carrier and its capture (recombination or trapping) determine the net number of carriers at a given instant.

Consider the case of a nearly intrinsic semiconductor with an equilibrium number of electrons and holes in the conduction and valence bands, respectively. Let the probability of recombination at a recombination center be R and the density of such centers be N_T. For purposes of simplification, assume that the probability of an electron being captured depends only upon its velocity. Assuming a Boltzmann distribution for the electron velocities, and an average velocity, $\bar{v}$, along with the probability of capture of a single electron $\sigma(v)$ (or capture cross-section) which, for this case, must be a function of velocity, then

$$R = \sigma(v)\bar{v}N_T n \tag{11-51}$$

The capture probability determines the decay rate; this is given by the time rate of change of the number of carriers as

$$R = -\frac{\partial n}{\partial t} = -\sigma(v)\bar{v}N_T n \tag{11-52}$$

Assuming that N_T is not a function of time (steady state), integration gives

$$\ln \frac{n}{n_o} = -\sigma(v)\bar{v}N_T t$$

or

$$n = n_o \exp\left[-\sigma(v)\bar{v}N_T t\right] \quad (11\text{-}53)$$

Here, n_o is the equilibrium number of electrons at the instant recombination started. The lifetime, τ, may be defined as the time between captures. Using Equation 11-52, this is

$$\tau = \frac{n}{R} = \frac{1}{\sigma(v)\bar{v}N_T} \quad (11\text{-}54)$$

This is substituted into Equation 11-53 to give

$$n = n_o \exp(-t/\tau) \quad (11\text{-}55)$$

A similar expression may be obtained for the recombination of holes.

The operation of many semiconductor devices depends upon the lifetimes of excess carriers which are injected into them. The excess carriers, n'_e and n'_h, are the numbers of carriers which exceed the equilibrium number. Since it is difficult to unbalance the electrical neutrality in a conduction process $n'_e \simeq n'_h$. In many cases, the density of excess carriers injected into strongly extrinsic material is considerably less than that of the majority carriers. In n-type material $n'_e \simeq n'_h \ll n'_{eo}$, where n'_{eo} is the majority carrier density prior to injection. Assuming that the traps are deep, and using Equations 11-52 and 11-54, it may be approximated that

$$\frac{1}{n'_e}\frac{\partial n'_e}{\partial t} = \frac{1}{n'_h}\frac{\partial n'_h}{\partial t} = -\frac{1}{\tau_o} \quad (11\text{-}56)$$

where τ_o is the average time for recombination to take place. If the time that injection stops is taken as $t = 0$, and it is assumed that recombination takes place only in deep traps, then

$$n'_e = n'_h = n'_o \exp(-t/\tau_o) \quad (11\text{-}57)$$

in which n'_o is the density of the carriers at the instant injection ceased. If $n'_e \gg n_o$ in a strongly p-type material, the excess electrons are the minority carriers and τ_o is the minority carrier lifetime. Similarly, in a strongly n-type material, the excess holes are the minority carriers and τ_o is the lifetime of the holes.

The lifeime of high, injected carrier densities in nearly intrinsic materials also is of importance in semiconductor devices. In this case, considering both electrons and holes, using Equation 11-56,

$$\frac{1}{n'}\frac{\partial n'}{\partial t} = -\frac{1}{\tau_{eo} + \tau_{ho}} \quad (11\text{-}58)$$

where n' is the total number of excess carriers at time t and τ_{eo} and τ_{ho} are the average lifetimes of the injected electrons and holes, respectively. The integration of Equation 11-58 gives

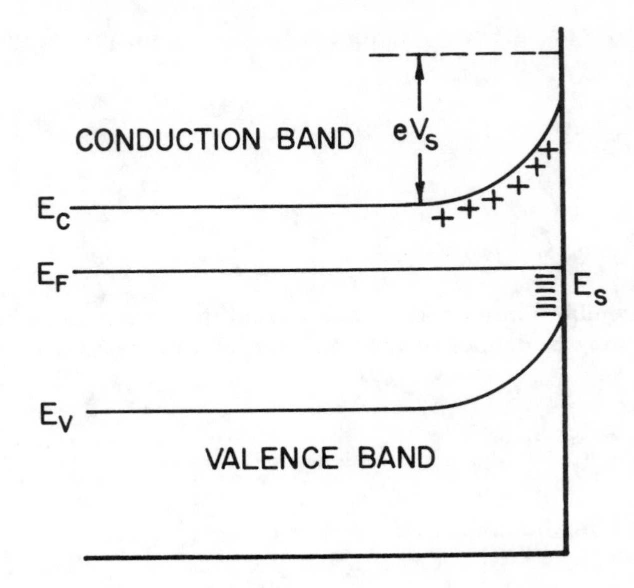

FIGURE 11-13. The effect of the surface space charge and the depletion zone upon the band structure of an n-type material. (Impurity states are neglected.)

$$n' = n'_o \exp\left[-t/(\tau_{eo} + \tau_{ho})\right] \tag{11-59}$$

This accounts for the differences in the lifetimes of both types of carriers.

The lifetimes of minority carriers are significant in semiconductor devices. This property also is important in the characterization of the quality of semiconductor crystals. Virtually all impurities and imperfections greatly reduce the carrier lifetimes. It is helpful that the most common impurities, C, N, and O, have almost no significant effects in Ge or Si. The theoretical values for lifetimes are of the order of seconds for direct recombination. The lifetimes in commercially produced materials range from about 10^{-6} sec to the order of 10^{-3} sec, but usually are about 10^{-5} sec.

11.2.1.1. Surface Effects

The preceding discussions relate to recombinations and lifetimes within the bulk of the semiconductor. The periodicity of the lattice is sufficiently uniform to provide a symmetrical band structure and carrier distribution within the bulk. The ions at or near the surface do not have the correct numbers of neighbors to satisfy their bonding. This leads to an asymmetrical charge distribution adjacent to the surface. The result is that localized states in the subsurface volume appear within the gap. These are known as Tamm states. They can act as traps because of the incomplete bonding. In addition, surface imperfections and impurities also constitute and contribute to these local, surface states. The asymmetric distributions adjacent to metal surfaces are very small because of the very large numbers of nearly free electrons and their random resonance between all of the ions in the lattice. In contrast to this, the asymmetry in semiconductors extends to a much greater extent into their volume than in the case of metals because their carrier densities are much lower than metals, even in extrinsic materials, and the covalent bonding.

This behavior is illustrated for an ideal, uncontaminated, n-type material. As drawn in Figure 11-13, these states are shown as acceptor states, E_s. The surface acceptor states are filled by electrons from the valence band. The resultant, thin, negatively charged, sub-surface layer acts as an electrostatic repulsion barrier for the electrons in the conduction band. The volume adjacent to this electrostatic barrier, thus, is virtually empty of electrons. This is known as a surface-depletion zone. This leaves positive donor ions and induces a positively charged volume next to the surface barrier. (Holes also may be attracted to surface states and produce an opposite effect.) The surface states fill up to the state at the same level as E_F and the potential barrier eV_s is induced. The surface charges cause the depletion zone to extend to about 10^3 ion diameters into the volume of the semiconductor. This volume tends to be more like p-type material. P-type material shows behavior opposite to that described here for n-type material.

The trapping of carriers at semiconductor surfaces is in addition to that which occurs in the bulk material. The lifetimes for surface trapping, or recombination, are very sensitive to the chemical and physical characteristics of the surface. Both surface impurities and imperfections are involved. Surface properties, consequently, are very important in the technology of manufacturing of semiconductor devices, especially those involving junctions. Suitable surface preparation and protection are required to prevent leakage or breakdown from occurring more quickly than would be the case within the bulk material. The inadvertent contamination by Na is one of the most common causes of this.

The surfaces of all semiconductor devices are contaminated to some degree; it is impossible to keep them chemically clean. The surface-impurity states also lie in the middle of the gap, contribute to the space-charge build-up and, thus, to the band distortions which are present adjacent to the surfaces. The normally high concentration of surface-impurity states causes E_F to lie approximately at the center of their distribution. The attachment of metal leads, or contacts, to the semiconductor surface results in only small displacements of the bands because the alignment of E_F takes place by the filling or emptying of the surface states. Metal-semiconductor rectification (Section 11.3) does not occur when the density of surface-impurity states is high. The surfaces of some semiconductors are prepared for such contacts by doping them so that the surface-impurity states either are completely filled or completely unoccupied so that rectification is avoided. Another method to accomplish this is to minimize the extent of the surface barrier (Section 11.5.1) so that tunneling takes place. Such nonrectifying metal-semiconductor contacts are called ohmic contacts.

Surface states can have important roles in metal oxide-semiconductor interactions (Section 11.10.2). Carefully prepared surfaces may contain fewer than 10^{10} impurity states/cm^2, while contaminated surfaces may contain more than 10^{13} impurity states/cm^2. Applied voltages can move E_F up or down because these states may fill or empty. These are known as fast surface states. Such states can cause variable, unstable responses in some devices because many of the surface contaminants, such as Na ions, can diffuse as a result of the applied voltages. Surface states also can contribute to breakdown (Section 11.6.1).

11.3. PHOTOCONDUCTION

Photoconduction is the increase in the electrical conductivity of semiconductors and some insulators as a result of irradiation by photons of sufficient energy. Photons with energies equal to or greater than E_g, as described in Section 11.2.1, are required to produce this effect. Carriers produced by photons with energies $E_g' < E_g$ (Figure 11-12) may be generated by mechanisms involving donor or acceptor states. In this phenomenon, the numbers of carriers created by phonons of energies E_a or E_d (Figure 11-9) are small and are neglected in the following discussion.

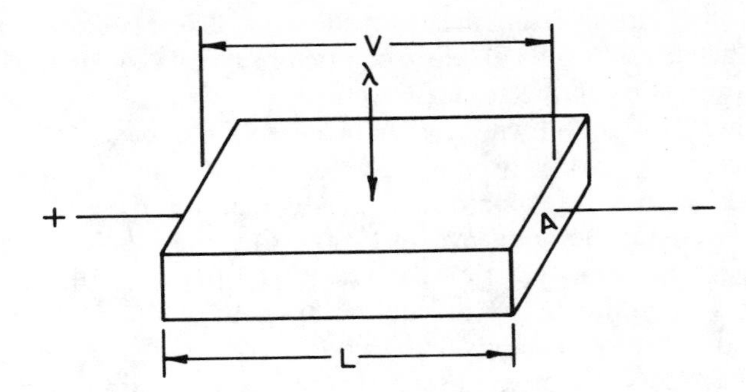

FIGURE 11-14. Simple model of a photoconducting crystal of section A and length L with a voltage V between its ohmic leads.

The maximum wavelength possible for electron excitation for a given intrinsic material is determined by E_g. And, since the wavelength associated with leV is $\lambda_o = 1.24 \times 10^{-4}$ cm, λ_o (microns) $= 1.24/E_g$; this is the intrinsic cut-off wavelength. In extrinsic materials, the wavelengths are longer and depend upon E_g' (Figure 11-9) or upon E_g'' (Figure 11-12). This is important in devices used for electromagnetic emission (see Section 11.8.2).

Consider a small, intrinsic single crystal which has a section area, A, exposed to incident electromagnetic radiation. The crystal is connected into an electrical, current-measuring circuit by means of ohmic leads as indicated in Figure 11-14.

A small voltage, and its corresponding electric field, will cause a small current to flow in the absence of radiation. This is the dark current; it results from the equilibrium number of thermally generated carriers. Since the relative number of carriers of this origin is small, these too will be neglected.

Large numbers of electron-hole pairs will be created when the incident photons have energies greater than E_g. This will cause a correspondingly large increase in the current. The holes have no effect upon the conduction process because of their very rapid association with recombination centers. If the total rate of generation of pairs in the crystal is g_r per second, the number of electrons flowing per unit volume per unit time in the crystal is g_r/AL. Using Equation 11-1, the conductivity is given by

$$\sigma_e = n_e e \mu_e = (g_r \tau_o / AL) e \mu_e \tag{11-60}$$

where τ_o is the average recombination time. Using Ohm's law

$$V = IR = I_e \rho_e \frac{L}{A}$$

and

$$\rho_e = \frac{AV}{I_e L} = [(g_r \tau_o / AL) e \mu_e)]^{-1} \tag{11-61}$$

This gives the steady-state current induced by the photons as

$$I_e = \frac{g_r \tau_o}{AL} e \mu_e \frac{AV}{L} = \frac{g_r \tau_o e \mu_e V}{L^2} \tag{11-62}$$

The lifetime is the most important factor in the generation of photon-induced currents (see Section 11.2.1.) The greater the crystal purity and perfection, the longer the lifetime and the larger the current which is generated. The current in crystals with long lifetimes, in which only the flow of the nearly free, photon-generated electrons is involved, is

$$I_e \simeq g_r e \tag{11-63}$$

because each electron which leaves the crystal is replaced by another from the circuit. By comparing Equations 11-63 and 11-62 the amplification, G, is

$$G = \tau_o \mu_e V/L^2 \tag{11-64}$$

Thus far, the trapping of the mobile electrons has been included in τ_o. This factor must be examined more closely to explain the steady-state behavior because both the current and the amplification depend upon the lifetime. Therefore, a better description of these electrical properties may be obtained if an approximation for the actual lifetime, rather than the average lifetime used above, was included. This may be done by means of Equation 11-55 rewritten as

$$n/n_o = \exp(-t/\tau); \quad \frac{n}{n_o} = \frac{1}{e^{t/\tau}}$$

Here, n_o is the number of electrons generated by photons in the absence of any trapping. This may be reexpressed, using the series approximation for the exponential term, as

$$\frac{n}{n_o} \simeq \frac{1}{1 + t/\tau}$$

Rearrangement gives

$$\frac{n_o}{n} - 1 \simeq \frac{t}{\tau}$$

or

$$\frac{n_o - n}{n} \simeq \frac{t}{\tau}$$

Then,

$$\tau = \frac{tn}{n_o - n} \tag{11-65}$$

The insertion of Equation 11-65 into Equations 11-62 and 11-64 results in

$$I_e = \frac{g_r e \mu_e V}{L^2} \cdot \frac{tn}{n_o - n} \tag{11-62a}$$

and

$$G = \mu_e V/L^2 \cdot \frac{tn}{n_o - n} \tag{11-64a}$$

The traps lying close to the bottom of the conduction band are very effective in the reduction of the number of carriers. At the beginning of illumination n is very small with respect to n_o; therefore τ and, consequently I_e and G are small. As the radiation

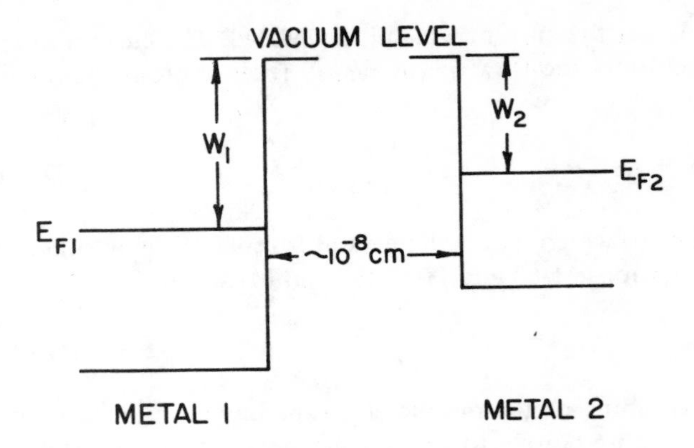

FIGURE 11-15. Potential wells of two metals, separated by a small distance, prior to interaction.

continues, the traps approach more complete filling so that n becomes larger with respect to n_o, and I_e and G increase correspondingly. This continues until all of the traps are filled. The traps remain filled as long as the illumination continues because they are not reactivated by holes. At times longer than that required to fill the traps, n becomes constant and the equations given above involving the average lifetime, τ_o, should be used. The traps have no further effect, once they are filled, as long as the radiation continues. The time required for a crystal to produce a steady current may be long (minutes) because of the time required to fill the traps. The traps may empty slowly when the illumination is turned off and the current may decay slowly, rather than abruptly.

11.4. METAL-METAL RECTIFICATION

Ordinarily at least one component of a rectifier is a semiconductor. However, this need not necessarily be the case, depending upon the work functions of the materials and the nature of their separation. Metal-metal rectification is presented here as an introduction to the more usual metal-semiconductor and semiconductor-semiconductor devices.

Consider two metals with work functions W_1 and W_2, respectively, so that they are separated by a distance of the order of 10^{-8} cm. The initial set of conditions is shown in Figure 11-15 at the instant that the two metals are brought together.

The wells are compared by placing their top levels at a common level. If the wells are close enough, electrons will travel from the second to the first metal by tunneling. The probability that an electron may be found outside of a potential well is shown as a function of well depth in Figure 3-3 (Volume I) (Also see Section 11.6.1). The electrons near E_{F2} will cross, or tunnel, from metal 2 to metal 1 when the potential barrier W_2 and the separation between the metals are small, particularly when a sufficient potential exists across the separation. In addition, some electrons may be thermally induced to leave well 2 and cross over to well 1 (thermionic emission). These processes leave the second metal positively charged and the first metal negatively charged. These mechanisms continue until the potential difference between the two metals vanishes and equilibrium is reached. At this point, the Fermi energies of the two metals are at the same level. The transfer of electrons now is the same in both directions and the result is two equal and opposite current densities, j_o, as shown in Figure 11-16.

Now apply a voltage, V, across the two metals which is smaller than the potential difference $(W_1-W_2)/e$. This causes the energy levels of metal 2 to become higher than

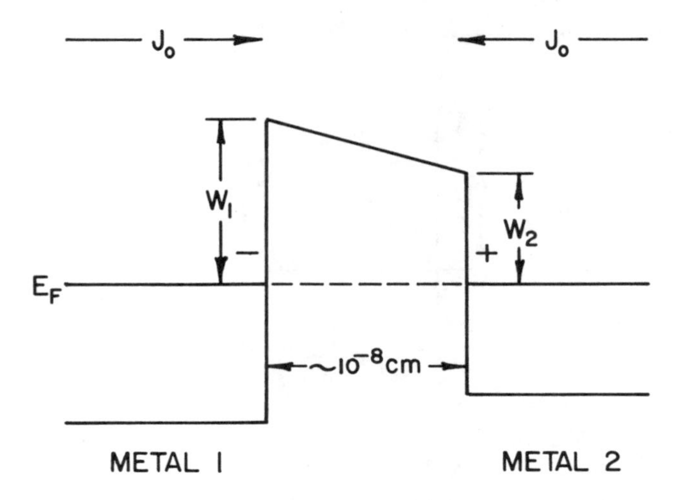

FIGURE 11-16. Equilibrium E_F and work functions of two metals separated by a small distance. (From Dekker, A. J., *Solid State Physics*, Prentice-Hall, Englewood Cliffs, N.J., 1957, 348. With permission.)

that of the equilibrium case by an amount equal to eV. The current density from the first metal to the second is still j_o. However, the rise in the energy levels of metal 2 by an amount eV decreases its potential barrier by the same amount (Figure 11-17). The decrease in the barrier increases the probability that an electron will cross the lowered barrier. Using the Boltzmann statistics, the increased probability of electron transfer from metal 2 to metal 1 is $\exp(eV/k_BT)$. The exponent is positive because V is negative. The current density from metal 2 to metal 1 now is $j_o\exp(eV/k_BT)$. The resultant current density caused by the forward bias is the difference between the two flows:

$$j_F = j_o\left[\exp(eV/k_BT) - 1\right] \tag{11-66}$$

In a similar manner, if a reverse bias is applied and metal 2 is positive to metal 1, the mechanism is reversed. The current density from the first to the second metal still is j_o, but that from the second metal to the first is $j_o\exp(-eV/k_BT)$. The resultant current becomes

$$j_R = j_o\left[1 - \exp(-eV/k_BT)\right] \tag{11-67}$$

In the forward direction, Equation 11-66 increases very rapidly with voltage, while in the reverse direction, Equation 11-67 rapidly approaches j_o; this is very small. Therefore, the direction in which the voltage is applied determines the magnitude of the current which will flow. The essentially one-way behavior of this kind permits rectification of current. This is shown in the "theoretical" curve of Figure 11-21. A metal-metal combination could provide a means for rectification, although this usually is accomplished by the means described in Sections 11.5 or 11.6.

11.5. METAL-SEMICONDUCTOR RECTIFICATION

Metal-semiconductor rectification depends upon the formation of an internal barrier layer at the interface between the two components. If the semiconductor is n-type,

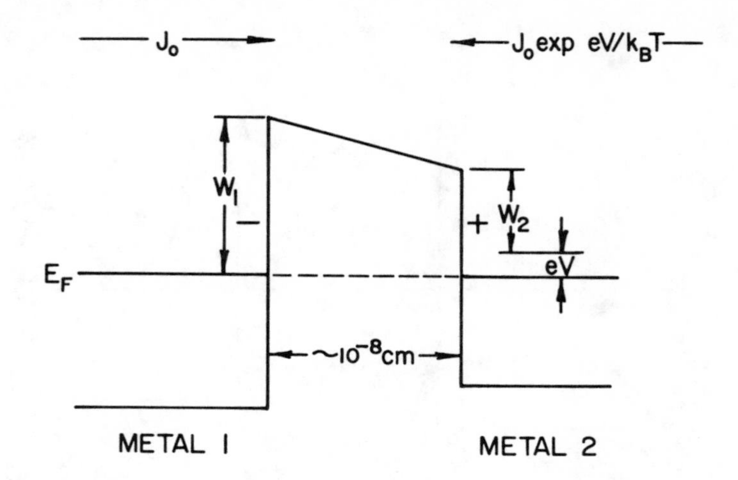

FIGURE 11-17. Potential barrier of two metals, separated by a small distance, when a voltage is applied so that metal 2 is negative to metal 1 (forward bias). (From Dekker, A. J., *Solid State Physics*, Prentice-Hall, Englewood Cliffs, N.J. 1957, 348. With permission).

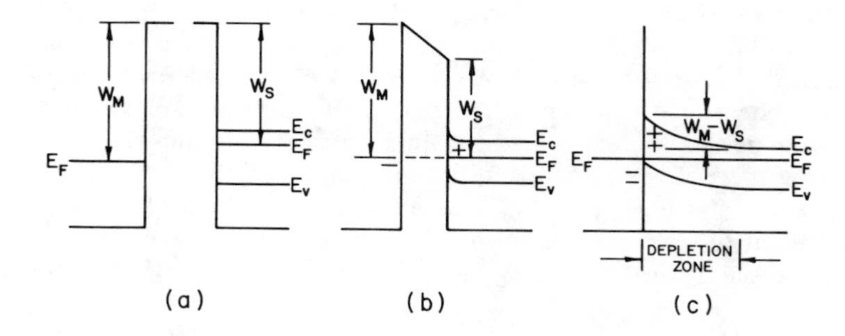

FIGURE 11-18. Metal-semiconductor (n-type) junction: (a) widely separated; (b) very narrowly separated; (c) after contact showing space charge and depletion zone.

electrons will diffuse into the metal from its conduction band and create positively ionized impurities. The result is a positive, space-charge, or depletion, region. The excess electrons in the metal induce a negative space charge. The two charged regions create an internal field in the interface region (Figure 11-18b). The mechanism is similar to that described in the previous section. The space-charge region within the semiconductor must lie within the range of the decrease of the potential difference, $(W_M - W_s)/e$ (the contact potential). At equilibrium, this potential difference results in the equalization of the Fermi levels and the numbers of holes and electrons in the barrier, or depletion, zone remain constant (Figure 11-18c). The larger the difference between the work functions, the greater will be the change in the bands of the semiconductor. This causes E_F to become closer to E_c. Barriers of this kind are known as Schottky barriers (see Section 11.9.6).

The diffusion of electrons from the metal to the semiconductor is similar to that described in the previous section for flow from metal 1 to metal 2, at equilibrium. This may be expressed as

$$n_e = n_{eo} \qquad (11\text{-}68)$$

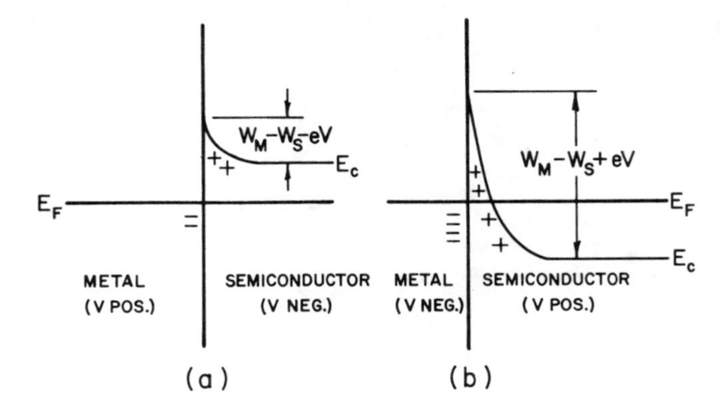

FIGURE 11-19. Effects of bias upon the bands at a metal-semiconductor junction (exaggerated for clarity): (a) forward bias; (b) reverse bias.

where n_{eo} is the equilibrium density of electrons. The application of a voltage, V, across the interface changes the bands of the semiconductor by an amount (−eV) (Figure 11-19a). As in the previous section, the number of electrons diffusing from the conduction band to the metal will be given by the general form

$$n_e = n_{eo} \exp(-eV/k_BT) \tag{11-69}$$

The net density of electron flow is the difference between Equations 11-68 and 11-69 for the forward direction, in general form, is

$$n_e = n_{eo} [\exp(-eV/k_BT) - 1] \tag{11-70}$$

In the same way, the net density of electrons for the reverse direction is, in general form

$$n_e = n_{eo} [1 - \exp(-eV/k_BT)] \tag{11-71}$$

The current density resulting from the electron flow is obtained from Equation 11-70, since $j = ne$, as

$$j_F = j_o [\exp(eV/k_BT) - 1] \tag{11-72}$$

When the voltage is applied so that the semiconductor is negative with respect to the metal (forward bias), the factor V in Equation 11-70 is negative, the barrier decreases, and the exponential term is positive; the current density will increase exponentially with V (Equation 11-72). When the bias is reversed, the barrier increases, the exponential term is negative (Equation 11-71), and a very small current will flow in the reverse direction. As in the previous section, the direction in which the voltage is applied determines whether a large or a very small current will flow; this permits rectification.

It would be expected that rectification should be affected by changes in $(W_M - W_s)/e$, the contact potential. In actuality such changes do not occur because of the local energy states within the gap at the junction. This results from behavior corresponding to that discussed in Section 11.2.1.1.

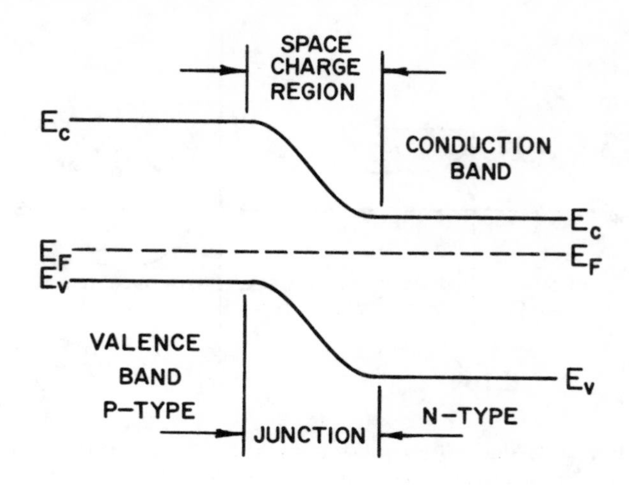

FIGURE 11-20. Band structure at a p-n junction.

Copper-cuprous oxide (CuO) and selenium oxide-normal metal diodes have been used for rectification; the former was discovered in 1926 and the latter in 1883, long before their operation was understood. They are capable of supplying relatively large d.c. currents and are used to operate motors, relays, battery chargers, and are essential components in d.c. power supplies.

11.6. P-N JUNCTION RECTIFICATION

Semiconductor crystals can be made so that they consist of both p-type and n-type regions which are separated by a thin, interfacial (junction) layer. This junction usually is of the order of 10^{-4} cm wide. While concentration gradients of both donor and acceptor ions exist across the boundary, (which to a large extent may be controlled), it is assumed for purposes of simplification that no such gradients exist. The simple band models for the individual, bulk p- and n-type materials are as shown in Figure 11-9. The band structures undergo a transition from that of one type to that of the other type of material at the junction. In some cases, the extent of the band transition must be kept small with respect to the mean-free path of a carrier of average lifetime (see Section 11.6.1). The electrons and holes will diffuse to the opposite sides of the junction, in the same manner as described in the previous section for electrons. The electrons diffuse into the p-type material and become minority carriers, creating positively charged donor ions by their absence. Holes diffuse into the n-type material and become minority carriers, creating negatively charged acceptor ions by their absence. The space-charge zone will attract all minority carriers within a volume which extends one mean-free path of a carrier on either side of it. At equilibrium two oppositely charged volumes build up and a strong internal field is created in the carrier-depleted volume about the junction (Figure 11-20). The effects of the space charge, in the depletion zone, and its behavior are discussed more fully in Section 11.9.6.

In the absence of an applied electric field, at equilibrium, an electron diffusion will occur from the n- to the p-type which will equal that of the opposite direction. This is analogous to Equation 11-68 and is denoted by

$$n_{-e} = n_{eo} = n_{+e} \tag{11-68a}$$

And a similar condition exists for the holes diffusing from the p- to the n-type material giving

$$n_{+h} = n_{ho} = n_{-h} \tag{11-68b}$$

The application of an electric field across the junction gives, for the electrons,

$$n_{-e} = n_{eo}\exp(-eV/k_BT) \tag{11-69a}$$

and

$$n_{+h} = n_{ho}\exp(-eV/k_BT) \tag{11-69b}$$

for the holes. The net density of the electron flow is

$$n_e = n_{-e} - n_{+e} = n_{eo}\left[\exp(-eV/k_BT) - 1\right] \tag{11-70a}$$

And that of the holes is

$$n_h = n_{+h} - n_{-h} = n_{ho}\left[1 - \exp(-eV/k_BT)\right] \tag{11-70b}$$

The net number of carriers is $n_e - n_h$. This is given by

$$n = n_e - n_h = (n_{eo} - n_{ho})\left[\exp(-eV/k_BT) - 1\right] \tag{11-73}$$

Then the current density is

$$j = ne = e(n_{eo} - n_{ho})\left[\exp(-eV/k_BT) - 1\right] \tag{11-74a}$$

or,

$$j = (j_{eo} - j_{ho})\left[\exp(-eV/k_BT) - 1\right] \tag{11-74b}$$

For forward bias

$$j_F = (j_{eo} - j_{ho})\left[\exp(eV/k_BT) - 1\right] \tag{11-75a}$$

where, as before, the current increases exponentially with the voltage, j_{ho} being very small. In the case of reverse bias

$$j_R = (j_{eo} - j_{ho})\left[\exp(-eV/k_BT) - 1\right] \tag{11-75b}$$

The negative voltage causes the exponential term to become very small and $j_R \simeq j_{ho} - j_{eo}$, which is very small. This behavior is essentially the same as that described for metal-metal and metal-semiconductor rectifiers. This is shown schematically by the broken curve in Figure 11-21.

The rectification capability of a p-n junction is apparent, provided that breakdown (Section 11.6.1) is not permitted to take place. Characteristics of these kinds are the bases for transistors and other junction devices. It will be noted that the forward portion of the curve for a realistic junction device will show a threshold voltage beyond which the j_F becomes appreciable. A typical value for the threshold voltage is about 0.5 V.

11.6.1. Breakdown

The curve in Figure 11-21 shows that j_R is negligible, in agreement with Equation

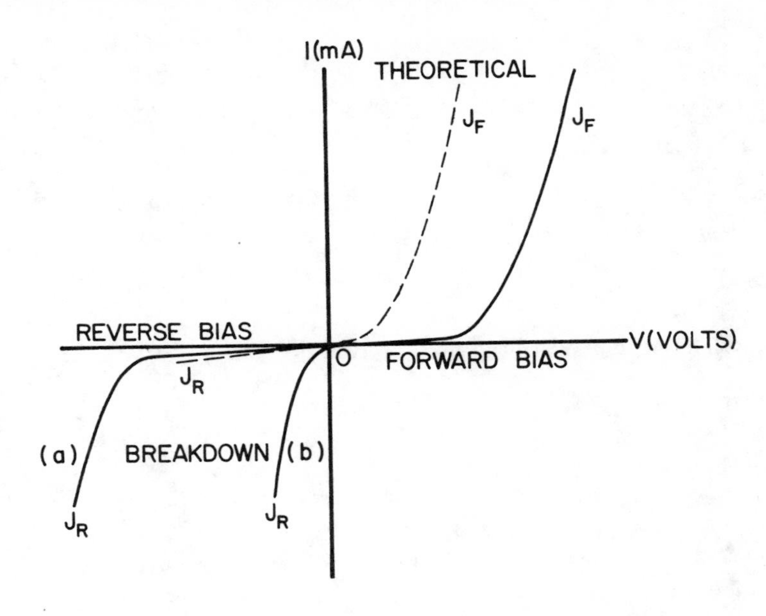

FIGURE 11-21. Schematic rectification behavior. Broken curve shows the theoretical behavior. Note the threshold voltage of a p-n junction under forward bias. (a) Avalanche and (b) tunnel breakdown.

11-75 up to a given voltage. Beyond this, j_R shows a very large increase which is known as breakdown. Breakdown which occurs at voltages of about 1 V or less is known as Zener breakdown and is caused by tunneling. Breakdown which occurs at higher voltages occurs as a result of pair production and is called avalanche breakdown. Junction devices are purposely made which utilize these responses.

When n- and p-type materials are heavily doped (at least $10^{19}/cm^3$), large numbers of electrons, or holes, fill the lowest possible energy states. These are degenerate states and E_F lies within the conduction and valence bands. Under these conditions, if the junction is thin enough (approximately 50 to 100 Å), the internal field within the junction will be at least 10^6 V/cm and the carriers may pass between the bands in the absence of an external voltage. No additional energy is necessary because E_F lies at approximately the same level with respect to each band. Thus, an electron near E_F in either band can cross the junction with no change in energy. This explains the extremely short response times and "internal breakdown" (Section 11.9.2) of devices in which tunneling is employed. The junction must be very small for tunneling in order that the electron wave functions can extend across the gap, since states are forbidden in the gap. A simplified diagram is shown in Figure 11-22.

In heavily doped materials with narrow junctions (because of the steep concentration gradients), in which E_F lies very close to E_c and E_v, but is located between these two, the application of a small, reverse bias of the order of 1 V, or less, is sufficient to shift the bands into the configuration shown in Figure 11-22. This permits large numbers of carriers to flow and explains those breakdowns which occur at very small voltages. Such necessarily small junctions can occur only when the dopant concentrations are high enough to cause very sharp concentration gradients at the junction.

Not all junctions are capable of breakdown by the tunnel mechanism. When the doping is such that the valence bands are insufficiently filled and the conduction bands contain too many electrons (cases in which the doping is high but is insufficiently concentrated) the junction width is large. This situation cannot provide the required

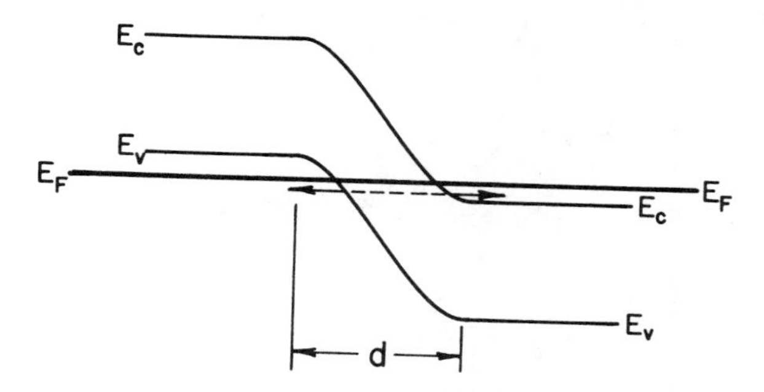

FIGURE 11-22. Direct tunneling across a very small junction of width d.

band relationships indicated in Figure 11-22. In addition, the large junction width will prevent tunneling because the wave functions cannot penetrate it and the internal fields are too small for tunneling.

Breakdown by the avalanche mechanism is considerably different from that induced by tunneling. When a sufficiently large reverse voltage is applied to the nearly free carriers in a semiconductor, a fraction of the carriers will have energies greater than E_g. A carrier with this energy may collide with an electron in the valence band, promote it to the conduction band and simultaneously create a hole in the valence band. Thus, much of the energy of the initial, high-energy carrier is consumed in the creation of two additional carriers. The initially excited carrier will have momentum and energy, E_o, equal to mv and $1/2\ mv^2$, respectively. After the reaction, assuming that the masses of the particles are the same, the momentum and energy of each will be 1/3 mv and $1/2\ m\ (v/3)^2$, respectively. The energy, E_R, which may be utilized in this reaction is

$$E_R = 1/2\ mv^2 - 3[1/2\ m(v/3)^2] = mv^2/3$$

So that

$$E_R = 2/3\ E_o \text{ and } E_o = 3/2\ E_R$$

The minimum energy for pair production must be greater than E_g, so that the minimum energy is $E_o = 3/2\ E_g$. This mechanism will be recognized as the inverse of the Auger process described in Section 11.2.1.

The avalanche mechanism is explained by considering n_o electrons which travel a given distance in a volume. An appreciable fraction, f, of these electrons will have energies greater than E_g if the voltage is large enough; these will create new pairs while traveling the given distance. The initial group of electrons will create n_of electrons en route to give a total of $n_o(1 + f)$ electrons. Simultaneously, n_of holes will have been created. Since the ionization energies of electrons and holes are of the same order of magnitude, it may be assumed that the holes behave the same as the electrons. Therefore, f holes also will generate pairs. This will result in the creation of n_of^2 electrons. In their turn, these electrons also will create additional pairs: n_of^3, etc. The total number of electrons generated in the volume is obtained by summing the geometric series, where $0 < f < 1$, as

$$n = n_o(1 + f + f^2 + f^3 + + +) = n_o/(1 - f)$$

The fraction f approaches unity for large volumes and for appropriate voltages. These

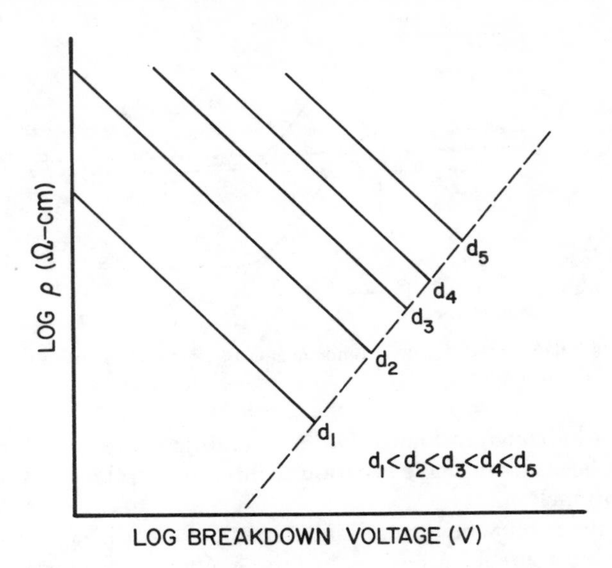

FIGURE 11-23. Schematic relationship of breakdown voltages with respect to junction width, d.

conditions cause a very large increase in the numbers of carriers. The origin of the name for this mechanism, and the way in which it causes breakdown, is apparent.

Surface states also contribute independently to breakdown. This arises from the fact that the electric field which is at the surface of a semiconductor is higher than that in the volume of the material. The depletion layer adjacent to the surface is smaller than that in the volume of the junction (see Section 11.2.1.1). Thus, breakdown at the surface occurs at lower voltages than in the bulk.

Avalanche breakdown occurs at relatively high voltages in moderately doped materials with comparatively large junctions. The breakdown voltage is a function of the concentration gradients in the junction; it may be as low as a few volts. It is responsible for breakdown in most transistors. In contrast, tunneling breakdown only can occur in highly doped materials with very small junctions; the breakdown voltages may be negligibly small. Some diodes are in reverse, internal breakdown, as a result of the high degree of doping, in the absence of any external voltage (Section 11.9.1). Both of these mechanisms may be operative in a given semiconductor device (see Section 11.9.2).

In the case of moderately doped transistors with relatively clean surfaces, breakdown will occur when the applied field is equal to or greater than the breakdown voltage of the depletion zone. The strength of the applied field increases as the width of the depletion zone decreases; $\overline{E} = V/d$, where d is the width of the depletion zone. As discussed here, and in Sections 11.9.2 and 11.9.6., d decreases as the dopant concentration increases. It is shown later (Equation 11-77a) that $d \propto (1/N_a)^{1/2}$ and that $d \propto V^{1/2}$. Thus, V increases more rapidly than d and the occurrence of breakdown becomes more likely as these variables increase.

The breakdown voltage varies directly as the width of the energy gap; the ionization energy of a material with a wide gap is greater than that of one with a smaller gap. (See Section 10.6 for an approximation for normal metals.) This is illustrated by the electric fields required for the breakdown of the following materials: Si, 3.0; Ge, 0.8; GaAs, 3.5; and SiO_2, 60 (all in units of $V/cm \times 10^5$).

The breakdown voltages of n-type silicon are shown schematically as a function of d in Figure 11-23.

11.7. SEMICONDUCTOR MATERIALS

A brief review of semicondutor materials is given here for convenience. The elemental semiconductors belong to group IV of the Periodic Table. These materials are insulators at 0 K (Chapter 5, Volume I, and Section 11.1). They become semiconductors as increasing temperature, or radiation, promotes electrons across the gaps to their conduction bands. Some values of the energy gaps of semiconductor materials are given in Table 11-2.

Diamond has the largest energy gap and is an insulator below about 1000 K. It shows intrinsic behavior above this temperature. The most widely used elemental semiconductor materials are Si and Ge. Si shows intrinsic behavior at about 200°C. Ge becomes intrinsic at about 100°C. Each of these two elements are virtual insulators below the temperatures given. The reason for the differing behaviors is that the values of E_g of these elements are significantly different. Diamond has the largest E_g and Ge has the smallest; that of Si lies between these, but is closer to that of Ge than to that of diamond. The electrical properties of these elements at temperatures below those given are determined by the impurities present (see Section 11.2.).

The difference between the energy gaps of Si and Ge determines that intrinsic Ge will show a greater conductivity than Si. More carriers, electrons and holes, will be available for conduction in the Ge than in the Si at a given temperature because of the smaller energy gap between its valence and conduction bands. However, this difference is of practical use. The preparation of intrinsic Si is less difficult than that for Ge since more impurity ions can be tolerated in states within E_g, because of its wider energy gap, before it shows extrinsic behavior.

A large number of compounds also demonstrate semiconductor properties. These usually have an electron:ion ratio of four (Table 11-2 and Section 10.6.7). Such compounds are covalently bonded and frequently crystallize in the zinc-blende and wurtzite lattices; they tend to have relatively large values for E_g. This makes it possible to dope them for many applications in devices including those involving junctions (see Sections 11.2 and 11.6).

Quite a few of the compound semiconductors show higher carrier mobilities than the elemental materials. This is useful because a given concentration of carriers will provide a higher conductivity in these compounds than in the elemental materials. It also will be noted that the materials cited in Table 11-2 vary largely in their values of E_g. They are useful beyond the range provided by the elemental materials because they permit more latitude in the selection of materials for use in the application of various devices.

Some of the compounds may have varying degrees of covalent bonding (Section 10.6.8). The degree of covalent bonding determines their applicability to semiconductor devices. In other words, the greater the amount of ionic bonding, the greater will be their tendency toward ionic behavior; such compounds tend to act as insulators.

The electron:ion ratio of four is not always required. Some compounds belong to IV-VI, II-IV, and II-V types (Table 11-2). Other compounds, not shown in the table, are of the V-VI, V-VIII, and III-VI types. These generally, but not always, have narrow energy gaps; many of the V-VI compounds have values of E_g from about 1.0 to 1.6 eV. Those with the smaller energy gaps are useful for devices which operate in the infrared region (Sections 11.3 and 11.8.1).

Two important factors must be given consideration in compound semiconductors. The first is that the stoichiometric ratios of the ions must be maintained as exactly as possible. The second is that their purity must be extremely high. Variations in the ratios of the ions would have the same effects as the intentional doping of the compound; it could be either n- or p-type depending upon which ion was in excess of the desired ratio. Impurities present in such compounds also act as donors and acceptors and cause extrinsic behavior where intrinsic properties may be desired. These factors make it more difficult to manufacture single crystals of intrinsic, compound semiconductors than elemental semiconductors. However these factors make it possible to produce extrinsic, compound semiconductors of either type by adjusting the relative amounts of constituent ions, without the introduction of dopants.

Metal oxides can show semiconductor properties when an excess of the metal ion is present. Zinc oxide can show n-type properties when this situation exists. The electrical conductivity of these crystals may be varied by controlling the amount of excess cation.

Defect oxides (Sections 10.6.7 and 10.6.9) also show semiconductor properties. These are nonstoichiometric compounds with deficiencies of either ion. Their lattices must contain vacancies. They may be considered to be semiconductors containing g acceptor or donor ions within relatively wide gaps. These contribute carriers and the oxide behaves as an extrinsic semiconductor.

Extrinsic semiconductor properties also are shown by defect semiconductors when the cation can have more than one state of ionization. Since the cation is present in the lattice with more than one valence, either cation or anion vacancies must be present in the crystal in order to preserve the electrical neutrality. The electrons may migrate between the cations with different valences. In some cases, the vacanies may permit ionic conductivity to contribute to that of the usual carriers. Metal oxides in which the cation has only a single ionization state, and in which there is no excess of cations, act as insulators.

11.8. BULK SEMICONDUCTOR DEVICES

The bulk properties of semiconductor materials form the bases for a number of simple, but very useful devices. It frequently is possible to make these from polycrystalline materials and many are made in this form. However, devices made from single crystals are more efficient since grain boundaries, which can act as traps and also decrease the mobility, are absent. The choice between the use of polycrystalline and single-crystal materials often depends upon the balancing of costs and properties.

The bulk devices included in the following sections are intended to provide further insight into the physical mechanisms involved in semiconductors and to serve as engineering applications of their properties. These are representative and not to be considered as inclusive.

11.8.1. Photoconductors

Suitable radiation incident upon a semiconductor can increase the number of carriers, and thereby increase its electrical conductivity (Sections 11.2.1 and 11.3). The intensity of the incident radiation determines the amount of additional current which flows. Low-intensity illumination can result in times of the order of minutes for the crystal to produce a steady current (Equation 11-62a); higher intensities can reduce this time to the order of seconds. This is caused by the fact that the rate of generation of electron-hole pairs, g_r, is affected by the intensity of the radiation. This also changes I_e (Equation 11-62) and the additional amount of steady flow, therefore, is a measure of the intensity of the radiant energy. Devices make use of this principle to detect and measure radiation.

The materials generally used for photoconducting devices are virtually insulators in the absence of radiation; the current is almost entirely a result of the incident, radiant energy. These materials are used because this application requires that they have moderately large values of E_g; this avoids the excessive introduction of thermally induced carriers and the resulting large, dark currents. E_g also determines the response of the semiconductor to radiation. The energy of a photon is inversely proportional to its wavelength. Therefore, λ must be sufficiently short in order that the incident photon may have enough energy to promote a valence electron across the gap to the conduction band. Thus, E_g determines the upper wavelength limit to which the semiconductor will respond (Section 11.3).

Materials which give photoelectric responses include B, C, Si, Ge, P, As, Se, Te, the sulfides, selenides, and tellurides of Zn, Cd, and Hg, alkali halides, oxides such as Cu_2O, MgO, BaO, and ZnO, as well as such III-V compounds as InSb, InSe, InTe, and GaAs. The more commonly used materials include CdS, CdSe, and CdTe. These have values of E_g near 2 eV. Leck* gives some "realistic" values for a practical CdS cell as $\tau_o \simeq 10^{-13}$ sec, $\mu_e \simeq 10^{-2}$ m^2/Vsec, V $\simeq$ 100 V and L $\simeq$ 1 mm. These give an amplification Equation 11-64 of about 10^3.

The response to irradiation decreases as λ becomes much shorter than the energy equivalent of E_g. The short radiation causes current to flow near the surface of the material, while most of it is absorbed by the bulk of the semiconductor without pair production. This decreases the conductivity. Therefore, the best response is obtained where the energy of the incident photons reasonably matches E_g.

Intrinsic materials are most effective for the detection of radiation in the near-infrared range. Far-infrared detectors make use of extrinsic materials (Si and Ge doped with group III and group V elements) in which the impurity levels are readily promoted by the long wavelengths. Mixed compounds such as HgTe-CdTe and Mg_2Pb-Mg_2Sn also have been used for this, but the compositions are difficult to control.

CdS crystals also have been used to detect high-energy gamma rays. A photon of about 1 MeV can create approximately 10^5 carriers; it is an excellent sensor for this kind of radiation. Materials for infrared detection must have smaller values of E_g. This may complicate their application because more thermally induced carriers may be present and the dark current can be relatively large. This may be reduced by cryogenic cooling. Compounds such as PbS, PbSe, and PbTe have been used for infrared detection without cooling.

Other applications include light meters and television cameras. Low-cost, thin-film, CdS solar panels for power generation now are available. Their efficiency is low (about 100 W/m^2), but their cost (about \$1.12/m^2) is more than two orders of magnitude less than those using silicon. So, despite their low efficiency, the savings are large and their application is practical (see Section 11.9.3).

Some other photoconductor applications are based upon the interruption of a beam of light. Such crystals must have short decay times. These are employed as counters, door openers, etc. They also are used to turn lights on when the illumination drops below a given level. Crystals which operate in the infrared portion of the spectrum have been used for the detection of aircraft and missiles, burglar alarms, etc.

Invisible images may be formed upon some of these light-sensitive materials. These are produced by photons absorbed by semiconductors or insulators of this class. A fine powder added to the surface will be attracted to the charge pattern and form an image. This is the basis for xerography. In this process, a metal surface is coated with a very thin film of Se. This film is charged electrostatically. The photons from the object to be photographed are transmitted to the Se film. The resulting photoconduction in those areas exposed to the light causes the charge to "bleed" away; the dark areas remain charged. A fine, thermosetting powder is spread upon the Se film and is attracted to the charged areas. The resulting pattern is heated, after being shifted to a piece of paper, and a permanent image results.

* Reprinted with permission from Leck, J. H., *Theory of Semiconductor Junction Devices,* Pergamon Press, Elmsford, N.Y., 1967.

Another process accomplishes the same objective, in essentially the same way as that just described, by the direct use of ZnO-coated paper.

11.8.2. Spontaneous and Stimulated Electromagnetic Emission

The light spontaneously emitted from a crystal is called luminescence. This usually is induced by ultraviolet or electron irradiation. Two types of responses occur. The case in which the emission stops virtually as soon as the irradiation is turned off is known as fluorescence. Here, the light emission stops within about 10^{-8} sec after irradiation stops. The other response is one in which the spontaneous emission of light can continue for long times after irradiation stops; it is known as phosphorescence and can vary from about 10^{-7} sec to hours.

Another type of electromagnetic radiation from materials is known as stimulated emission. It results from the forced population of ionized states of impurity ions in solution. The resultant radiation is coherent in contrast to luminsecence and phosphorescence which emit incoherent radiation.

11.8.2.1 Photoluminescence (Phosphorescence)

High-energy, incident photons can promote many electrons from the valence to the conduction band. The electron-hole pairs move through their respective bands until they are trapped (Figure 11-12). The electrons may jump back and forth between traps and the conduction band, with the rate of escape given by $n(\nu)\exp(-E_t/k_BT)$ (see Section 11.2.1). The type of impurity and its degree of ionization determine E_t. The factors E_t, T, and the number of traps determine the number of times an electron jumps back and forth between the traps and the conduction band before it drops down into a luminescent center and emits its extra energy as a photon. This explains the long emission times shown by some phosphorescent materials. Materials in which this mechanism occurs over short times are short-persistence phosphors; those lasting longer times are long-persistence phosphors. Both are types of luminescent materials.

An electron dropping from the conduction to the valence band via the process shown in Figure 11-12c may give up its extra energy as a photon. This is a radiative transition known as spontaneous emission. The photon emitted by the path indicated in Figure 11-12d would occur at a longer time and have a longer wavelength than that shown in Figure 11-12c. A subsequent drop to the conduction band may be nonradiative and phonons rather than photons may be produced. The emission times are determined by the number and degree of ionization of the coactivators (traps just below the conduction band), while the spectrum of emitted photons is determined by the degree of ionization of the activators (recombination centers or traps above E_v). The maximum wavelength possible for a given intrinsic material was shown to be λ_o (microns) $= 1.24/E_g$, the cut-off wavelength (Section 11.3). Longer emitted wavelengths, which depend upon the energy differences, E_g'', between the activators and coactivators, account for the spectrum emitted by a given material.

Another mechanism is based upon the presence of impurity ions. These ions in either ionic or covalent solids may have both excited and ground states which depend upon the surrounding ions in the lattice (Figure 11-24). The excited state may be the result of the capture of an electron from the conduction band, or by the exchange of energy with an exciton. (Excitons are produced by photons with $h\nu < E_g$. They are electron-hole pairs which are electrostatically bonded because the incident photon had insufficient energy for their complete separation. The energy for exciton production is about $0.7\ E_g < h\nu < 0.9\ E_g$. Excitons are electrically neutral, but can transmit energy because of their mobility.) An excited electron will return to the ground state if it emits a photon. Such a photon can have energies which vary between $h\nu_1$ and $h\nu_2$, depending upon the position of the oscillating ion at the instant of emission of the photon. This

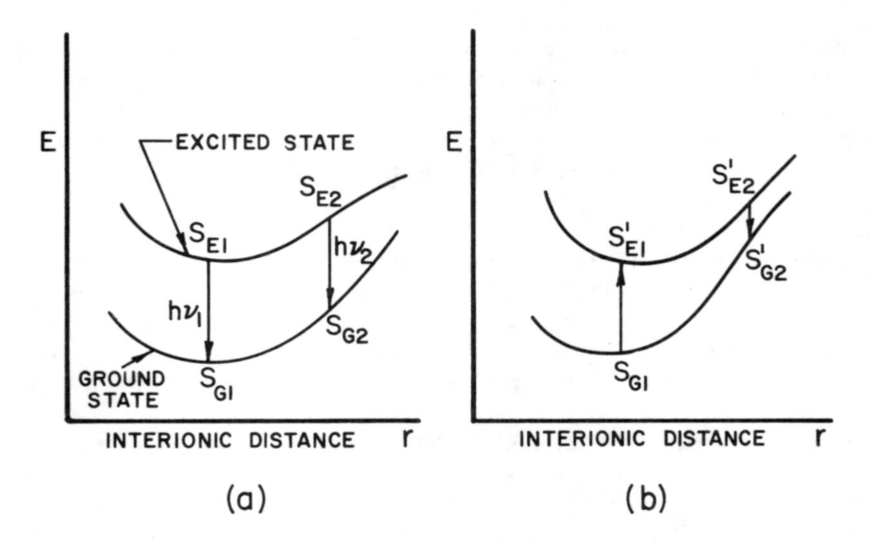

FIGURE 11-24. (a) Radiative transitions. Excited and ground states of an excited impurity ion. S_{G1} and S_{G2} represent ground states and S_{E1} and S_{E2} are corresponding excited states. (b) Nonradiative transitions are indicated by primes.

results in the emission of light over a range, or band, of frequencies. The electron transition increases the vibrational amplitude of the impurity ion; it will be at a higher energy after $h\nu_2$ has been emitted than after $h\nu_1$. The energy difference between the two transitions appears as a phonon with energy $\Delta E = h(\nu_1 - \nu_2)$.

The color range of the emitted photons depends upon the frequency range between ν_1 and ν_2. This depends upon the ionization energies of the impurities present. The change in energy associated with this frequency difference is uncertain because of the uncertainty of the time that the photon will be emitted (Equation 2-35, Volume I). This energy uncertainty gives the natural line width of the emitted light.

Nonradiative transitions may occur in which only phonons are produced; they may result from interactions induced by phonons. In other cases, photons in the infrared region of the spectrum are produced. Impurities reacting in these ways are undesirable. Such impurities have been termed "killers."

These reactions can be explained by a mechanism suggested by Seitz, with the help of Figure 11-24b. After excitation, the equilibrium lattice position of the excited state may shift from S_{E1} to S_{E1}'. The emission of a photon in the infrared range would cause it to drop across the narrow energy interval to S_{G2}'. The remaining energy would be given up to the lattice as phonons. Phonons also could cause transitions by means of a path such as $S_{G1}S_{E1}'S_{E2}'S_{G2}'S_{G1}$ in which the excess energy is imparted to the lattice.

The main engineering applications of luminescence is in "fluorescent" lamps and in television and oscilloscope tubes. Since most phosphors do not luminesce in the pure state, impurity ions are added for this purpose (Figure 11-24a). Materials used for this purpose include ZnS with Mn impurities, $CaSiO_3$ with Pb and Mn impurities, and $Ca_3(PO)_4$ with Mn and Ce impurities. The low-pressure argon and mercury-vapor discharge in a fluorescent lamp produces ultraviolet light. This reacts with the impure phosphors on the inner surface of the lamp and produces emissions in the visible portion of the spectrum. The luminescence generated in this way occurs without the necessity for heating and comparatively little energy is dissipated. Incandescent illumination, in contrast, requires very high temperatures and converts very little energy into visible radiation.

The electron beam within a television tube reacts with phosphors upon the inner face of the tube and produces light. The picture is obtained by variations of the intensity of the electron beam as it sweeps across the tube. Phosphors used for this purpose are required to have moderate persistence so that rapid motion can be reproduced without the formation of ''ghosts'' or of flicker.

11.8.2.2. Stimulated Emission

The radiation described in the previous section was a result of spontaneous emission. A comparatively new, and increasingly useful, type of emitted electromagnetic radiation is generated by stimulation. Devices which generate this type of radiation are called masers and lasers (microwave, or light, amplification by stimulated emission of radiation). Inorganic and organic crystals, liquids, and gases have been used for these purposes.

When a photon with energy $h\upsilon$, with a natural line-width $\Delta\upsilon$, reacts with an excited ion whose energy above the ground state is exactly that of the incident photon, the reaction will stimulate the immediate emission of another photon. The emitted photon will have properties identical with that of the incident photon; it will have the same frequency, be in phase with, and have the same direction as the incident photon.

Under equilibrium conditions, states with lower energy normally are more highly populated than states at higher energies. The Boltzmann equation, (Section 5.4.1 Volume I), describes this situation as

$$N_e = N_o \exp(-E_e/k_B T)$$

in which N_e and N_o are the numbers of excited and ground-state ions, respectively, and E_e is the energy of an excited state. When excitation, or ''pumping'', occurs (by electrical discharges in gases, by microwave oscillation or by rapid optical flashing), the equilibrium conditions no longer exist and N_e becomes greater than N_o. This is called population inversion. Stimulated emission is possible only when such an inversion is produced. This must be accomplished very rapidly because the lifetime of an excited state is very short.

This process requires that the amplification produced by the stimulated emission be equal to the energy given up by the energy source. The excited ions are in a cavity resonator so that the oscillations resulting from the excited ions form a coherent standing wave. Each additional photon, given up by an excited ion, adds to the intensity of the standing wave and amplifies it. The cavity for use in the microwave region consists of a container whose length, L, is given by $n\lambda = 2L$, where n is an integer and λ is the wavelength of the emitted photons. The length of the cavity is made much greater than its diameter. This gives a single mode of axial microwave oscillation for each L. Many axial modes of oscillation are possible with λ in the optical region because of the much shorter wavelengths. The shorter the cavity, the fewer modes.

In an ideal cavity, one in which all of the applied energy produces excited states which generate coherent radiation, a stimulated ion spontaneously releases a photon of $h\upsilon_e$ and drop back to its ground state. A simultaneously emitted phonon reacts with the photon upon another excited ion. This produces another photon of $h\upsilon_e$ and the process cascades. All of the emitted photons have the same frequency, within the limits set by the Uncertainty Principle. This narrow frequency range is maintained because as one ion returns to the ground state, the phonon emission permits another stimulated ion to relax into the precise, equivalent, lattice vibrational state originally occupied by the first ion. These relaxation times are of the order of 10^{-12} sec. Thus, each emitting ion is at the same phonon frequency as all other emitting ions.

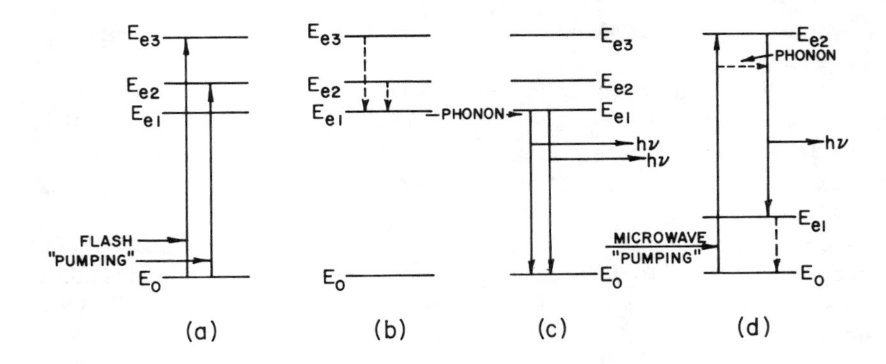

FIGURE 11-25. Ruby laser mechanism: (a) excitation; (b) spontaneous nonradiative transition to a metastable state; (c) stimulated return to the ground state and photon emission; (d) three-level maser.

The energy levels of the stimulated ions must either be very narrow or the gaps between the levels must be very distinct for this to occur. Situations other than this would be very inefficient, since only a few ions could react to the stimulating photons and phonons. This would result in low intensities and wide line widths. The natural line-width decreases, under the preferred conditions, as the temperature decreases because the number of phonon modes decreases and more ions may be stimulated for a given vibrational mode (Figure 4-3, Volume I).

A common method of cavity formation is to place two mirrors, either dielectric or metallic, at the ends of the material being stimulated. The ends of the cavities must be as nearly parallel to each other as possible and be optically flat to avoid destructive interference. One of the mirrors either contains a small opening or only is partly reflecting. Part of the standing wave induced in the masing or lasing material leaves it by these means. The largest part of the radiation remains within the cavity. The emergent beam of radiation is of high intensity ($> 10^4$ W/cm^2). It also has a very small degree of dispersion.

Solids, liquids, and gases may be stimulated to emit coherent radiation. The ruby maser was one of the first of these materials; it also is used as a laser. Synthetic ruby is Al_2O_3 doped with paramagnetic Cr^{+3} ions. These are in substitutional solid solution on Al^{+3} sites in the HCP lattice. The Cr^{+3} ion has three states of ionization in a magnetic field. Its levels may split into three levels, while O^{-2} and Al^{+3} ions each have single ionized states which primarily are involved in the covalent bonding. These ions also have low nuclear magnetic moments (Section 8.4 Volume II), and, thus, do not affect the levels of the Cr^{+3} ions. The very low concentration of Cr^{+3} ions assures that its levels will not be affected, or broadened, by the presence of other ions (see Section 8.6, Volume II). A schematic diagram of the band structure of the Cr^{+3} ion is shown in Figure 11-25.

The pumping photons usually are provided by intense, short pulses of light in the green portion of the spectrum (Figure 11-25a). A spontaneous, nonradiative transition (Figure 11-25b) results in drops to level E_{el}. Stimulated emission then causes the drop back to the ground state (Figure 11-25c). The photons emitted by these transitions are reflected many times within the laser. They are in phase with the standing wave in the cavity and, thus, account for the amplification by the laser. The high aspect ratio of the cavity minimizes transverse oscillations; other radiation not parallel to the crystal axis is lost and does not enter into the amplification process. The maser mechanism is the same as that for the laser. The only differences are the pumping by a microwave generator and the emission of amplified, coherent microwaves instead of light waves (Figure 11-25d).

Many suitably doped compounds have been used for masers. The dopant almost invariably is a paramagnetic lanthanide or actinide ion capable of suitable degrees of ionization. In addition to Al_2O_3, other host crystals include CaF_2, SrF_2, BaF_2, $SrWO_4$, and $CaMoO_4$ doped with such ions as Pr^{+3}, Nd^{+3}, Sm^{+2}, Dy^{+2}, Ho^{+3}, Er^{+3}, Yb^{+3}, and U^{+3}.

P-n junctions in semiconductors are the most efficient lasers. Coherent emission results from recombination adjacent to the line of the junction. Compounds such as GaAs, InAs, GaSb, InSb, and GaP are eployed. These are doped to degeneracy on both sides of the junction so that the energy gaps correspond to the desired frequency of the emitted radiation. The junction must be very narrow and extremely straight for the generation of the standing wave. Forward biased junctions may be pumped optically, by an electron beam or by the injection of electrons or holes. These III-V compound junctions usually are operated below 77 K (liquid N_2 temperature).

Applications for lasers now vary widely and are employed in such diverse ways as melting, welding, eye and neurosurgery, long-distance carrier waves for communications, and a host of others. Masers find wide use as amplifiers and carrier-wave generators for long-distance communications, especially when cryogenically cooled. Some of their more spectacular applications include radio astronomy and communications and telemetry in space exploration. Serious consideration is being given to their use in the transmission of solar energy from satellites (Sections 11.3 and 11.8.1).

11.8.3. Mechanical Devices

The properties of semiconductors respond to the application of mechanical stresses. Since these materials are not close-packed, their interionic distances change, comparatively easily, when mechanical stresses are applied. This changes the shapes of their Brillouin zones (Section 5.8, Volume I) and, consequently their band structures and energy gaps (Figure 11-7). A detailed analysis is too complex for this text. Simply stated, the stress tensors change the conductivity tensors (Section 11.1.2) which are functions of the crystallographic directions. The resultant change in conductivity is a piezoresistance effect. This was observed first by P. W. Bridgeman in metals (1925).

A simple, one-dimensional lattice is used here to provide an insight into piezoresistance. Under hydrostatic pressure, the interionic distances decrease and the Brillouin zone becomes larger; the energy gaps change (Figure 5-22c, Volume I). When uniaxial tension is applied to this lattice, the interionic distance increases, the extent of the Brillouin zone decreases, and the energy gaps again change. In real crystals the crystallographic orientation which gives the greatest change in resistivity per unit stress is the preferred orientation for a given semiconductor (see Figure 11-7).

The tensor-tensor analysis gives the uniaxial stress-gauge coefficient in terms of the tensor

$$R_e = \Delta R / R_o \sigma_x$$

where ΔR is the change in resistance, R_o is the initial resistance, and σ_x is the uniaxial stress. The stress-guage coefficient may be calculated from Ohm's law using the expression

$$\bar{E}/\rho_o = j(1 + R_e \sigma_x)$$

in which $\overline{E}$ is the electric field, ϱ_o is the initial resistivity, and j is the current density. The strain-gauge factor is obtained as

$$\gamma = E_y R_e = \frac{\sigma_x}{\epsilon_x} \cdot \frac{\Delta R}{R_o \sigma_x} = \frac{\Delta R}{R_o \epsilon_x}$$

Here E_y is Young's modulus and ε_x is the uniaxial strain.

Devices based upon this effect employ sensitive elements with high aspect ratios. This is done to avoid more complicated tensors than R_e. Thus, long, fine wires or comparable films or layers are used. The necessity for high values of γ has led to the use of n-type Ge and p-type Si. Both of these materials have high shear strengths in the [111] direction. Si devices oriented in this way have the highest values of γ; values up to 175 have been reported. These have a linear resistance-strain relationship. In using such gauges it is important that the temperature of the gauge be held reasonably constant. The large temperature coefficients of resistivity can introduce inaccuracies and cause erroneous strain readings.

Other devices based upon piezoresistance include pressure sensors, phonograph pick-ups, microphones, and accelerometers.

11.8.4. Temperature Measurement and Resistance Change

Intrinsic semiconductors have large, negative temperature coefficients of electrical resistivity (Figure 11-2). This sensitivity of resistivity to temperature changes is ideal for the measurements of temperatures. Extrinsic semiconductors usually have a similar, but smaller, response to temperature changes, but as will be shown below, some may have positive temperature coefficients at very low temperatures.

Starting with the reciprocal of Equation 11-1, and taking the derivative with respect to temperature, gives

$$\rho = (ne\mu)^{-1}$$

and

$$\frac{d\rho}{dT} = \frac{1}{e}\left[-\frac{1}{n\mu^2}\frac{d\mu}{dT} - \frac{1}{n^2\mu}\frac{dn}{dT}\right]$$

The temperature coefficient of electrical resistivity is obtained from these relationships as

$$\alpha = \frac{1}{\rho}\frac{d\rho}{dT} = ne\mu \cdot \frac{1}{e}\left[-\frac{1}{n\mu^2}\frac{d\mu}{dT} - \frac{1}{n^2\mu}\frac{dn}{dT}\right]$$

This simplifies to

$$\alpha = -\frac{1}{\mu}\frac{d\mu}{dT} - \frac{1}{n}\frac{dn}{dT}$$

Phonon scattering increases with increasing temperatures so that, except for very low temperatures, $d\mu/dT$ will be negative (see Section 11.8.5). The generation of carriers will increase with temperature, so dn/dT will be positive and larger than $d\mu/dT$. Therefore, α will be negative. In those extrinsic semiconductors in which the number of carriers is nearly constant, and varies only slightly with temperatures, dn/dT will be small and positive, while $d\mu/dT$, as before, will be negative. Thus, α will be positive for this set of conditions.

Bulk semiconductor devices whose utility and applications depend upon their changes in resistance with temperature are given the generic name "thermistors". Such devices usually consist of encapsulated, self-heated semiconductor materials with large temperature coefficients of electrical resistivity. Some of the important device parameters include the type of semiconductor material, the temperature coefficient in the temperature range of interest, the electrical resistance, heat capacity, thermal dissipation, and power.

The early thermistors were made of mixtures of manganese- and nickel oxides. A more recent material consists of NiO doped with Li_2O. The amount of Li_2O determines the number of Ni ions and permits variations in the electrical resistivity. Fe_2O_3 which contains prescribed amounts of "FeO" is changed from an insulator to a material with desirable properties for thermistors. Wustite normally is designated incorrectly as FeO because the ratio $O/Fe > 1$. The combination of the two oxides results in a defect structure in which the Fe ion has more than one ionization state. The results in semiconductor properties (see Section 11.7). Some extrinsic thermistor materials with positive temperature coefficients of electrical resistivity include suitably doped Si and such oxides as $BaTiO_3$ and $SrTiO_3$.

The obvious applications of thermistors include temperature measurement and control. Other applications include the compensation of the resistance of electrical circuits, vacuum gauges, and fire-alarm sensors.

11.8.5. Resistors

At moderate temperatures conduction occurs by both electrons and holes in intrinsic and extrinsic materials. The resistivity normally increases as the temperature is lowered because the rate of decrease of the number of carriers resulting from the filling of donor states, the emptying of acceptor states, and the return of electrons from the conduction to the valence band is greater than the increase in mobility (Section 11.8.4). In other words, the number of nearly free carriers diminishes faster than their mobility increases as the impurity levels and the valence band refill. The impurity scattering limits the mobility at very low temperatures. In extrinsic materials the hole contribution to the conductivity (Equation 11-3) can be larger than that due to electrons, even though their mobility is limited. The result is that the resistivity ceases to increase and shows only a small, positive slope with decreasing temperature. For example, Ge shows this behavior below 20 K, depending upon its purity.

Another mechanism may take place at low temperatures when the numbers of donors and acceptors are small, i.e., the material is nearly intrinsic. An ionized pair consisting of an acceptor and a donor ion can be considered as being analogous to an ionized hydrogen molecule, if they are close enough. The application of a voltage (external electric field) will cause the electron to jump to other impurity ions; a small degree of conductivity results from this mechanism.

Semiconductor resistors are used widely because they are reliable and inexpensive. These are available with either linear or nonlinear temperature coefficients (Section 11.8.4). Those with linear coefficients find use in the compensation of resistance in electrical circuitry. Those with nonlinear coefficients have been used for overload controls.

Another class of semiconductor resistors is voltage-sensitive and is marketed under many trades names such as Varistors® (Western Electric). These are made from sintered SiC particles and ceramic binders. Their properties may be varied by the use of proprietary additions and adjustments in compacting pressures, sintering times, and temperatures. Metallized contacts are provided. The adjacent SiC particles are considered to act as rectifiers. This enables them to behave as voltage-sensitive resistors. The voltage-dependence of the current which occurs in rectification permits them to act as though the resistance varies with the voltage. The SiC resistors are sturdy, have little reaction to overloading, and are not excessively expensive. P-n injunctions in Ge, Si, and GaAs also are used for this purpose.

Variable resistors of another kind may be made by combining a photoconductor (Sections 11.3 and 11.8.1) and a source of illumination. The illumination, from a bulb, passes through a fixed aperture before striking the photoconductor. The illumination is varied by voltage changes in the circuit controlling the filament. The same results may be obtained using constant illumination and employing mechanical changes in the opening of the aperture. Polycrystalline CdS usually is used for these devices. The dark current is small, the sensitivity to the visible portion of the spectrum is good, and the range of change in resistance is broad and reasonably stable, when properly encapsulated.

11.8.6. Thermoelectric Devices

The Thomson thermodynamic relationships, derived in Chapter 7 (Volume II) for metallic conductors, also hold for semiconductors. However, the absolute thermoelectric powers of semiconductors are much larger than those of metals. This has made their application in power generation and refrigeration possible.

The Peltier effect (Section 7.2, Volume II) is defined as

$$P_{AB} = \frac{\Delta Q}{I}$$

Equation 7-13 (Volume II) may be re-expressed as

$$P_{AB} = \frac{dE_{AB}}{dT} T$$

Equating these expressions gives

$$\Delta Q = I \frac{dE_{AB}}{dT} T$$

This relates the thermal and thermoelectric properties. Thus, where the A-B combination of thermoelements provides a large Peltier heat change, it also generates a large thermoelectric power. This means that thermoelements which are desirable for refrigeration also are good for power generation.

The emf generated by such a pair of thermoelements is derived from Equation 7-18a (Volume II) as

$$E_{AB} = \int_{T_1}^{T_2} (S_A - S_B)\, dT \equiv S_{AB}\Delta T$$

It will be recalled (Section 7.3, Volume II) that the Thomson heat generated is equal to $\sigma j dT/dx$ and the heat will be absorbed or liberated depending upon the direction of the flow of the current with respect to the temperature gradients in the thermoelements. As the temperature of the heated junction of two semiconductors is increased, the majority carriers in each leg will diffuse to the colder portions of the respective component. The most efficient thermoelectric circuit, thus, is provided by a thermocouple made of a pair consisting of n- and p-type elements (Figure 11-26).

The configuration shown in Figure 11-26 is that intended for the generation of power. Groups of thermocouples of this kind are connected in series and arranged so

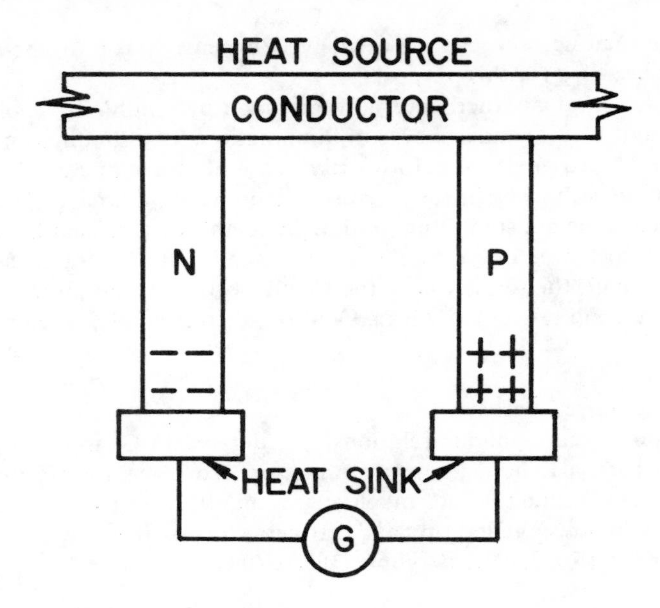

FIGURE 11-26. Schematic diagram of a portion of a thermoelectric generator. Conductors and heat sinks are metallic.

that their hotter junctions are in good thermal contact with a heat source such as that provided by nuclear fission. This technique has been used to provide power for satellites.

If, instead of creating a temperature difference between the junctions, current is made to flow in the same direction as the Seebeck current, heat will be absorbed at the former heated junction and will be liberated at the prior cooler junction. The absorption of heat by a series arrangement of thermocouples of this kind about a container constitutes a means of refrigeration.

In either of these applications, the electrical conductivity, σ, of the thermoelements plays an important role. The Joule heating, I^2R, represents extraneous heating in refrigeration and a loss of power in power generation. The thermal conductivity, $\varkappa$, allows heat to flow into the refrigerated zone and permits heat losses from the heated junction, decreasing ΔT in power generation. The most desirable materials for these purposes, therefore, should have high σ and low $\varkappa$. It will be recalled, from the Wiedemann-Franz ratio Equation 5-69, (Volume I), that these requirements are incompatible. Nevertheless, these two factors are fundamental in the prediction of the efficiency of thermoelectric devices for these purposes.

Present theories have not been successful in the prediction of the thermodynamic efficiency of semiconductor materials. Instead, a figure of merit, Z, is used to describe their properties for these applications. This is given by

$$Z = \frac{S^2}{\rho \kappa}$$

in which S is the absolute thermoelectric power. Other expressions for the figure of merit are available, but these are not as commonly used as the one given here.

The figure of merit is a function of the number of carriers. S decreases as the number of carriers, n, increases because $[\partial \varrho n \phi(E)/\partial E]$ in Equation 7-41 (Volume II) decreases as n becomes large. Both σ and $\varkappa$ increase with n. This results in a maximum for Z as a function of n which lies between 10^{18} to $10^{20}/cm^3$. Since n is a function of temperature, Z also is a function of temperature. Z usually lies in the range of about $1 \times 10^{-3}/$

deg to 3×10^{-3}/deg. Values in this range give thermal efficiencies of about 5%, although values as high as 13% have been reported. The problem is one of finding the best balance of S, σ and κ for a given application. Such a balance results in absolute thermoelectric powers in the neighborhood of 250 mV/°C.

Some materials with high values of Z, near 500°C, include $Pb_{1.05}Te_{0.95}$, Bi_2Te_3 and $Bi_2Te_{2.4}Se_{0.6}$; both Bi-base compounds are doped with CuBr. $CuGaTe_2$ has a value of $Z \simeq 3.0 \times 10^{-3}$ near 500°C. These are among the better materials for power generation. Bi_2Te_3 has applications for refrigeration devices. Refractory materials are used at temperatures higher than 1000°C; these have $Z \sim 10^{-4}$.

The low efficiencies of these devices make them acceptable only in remote locations or in unusual situations. These include power generation where conventional sources are unavailable, such as space vehicles or satellites. They constitute long-lived power sources in space applications, in which the thermal energy is provided by spontaneous nuclear fission. In addition, since no moving parts are involved, they do not affect the orbits of satellites.

Thermoelectric generators heated by nuclear reactors would be ideal for submarines, if sufficient power could be generated. Another limitation is the inefficient use of the nuclear fuel. Their replacement of diesel- or turbine-driven generators (with efficiencies in the neighborhood of 30 to 50%) would eliminate a major source of detection when submerged. Another limitation is the effect of neutron irradiation upon the semiconductor materials. A wide range of responses to such radiation is shown by this class of materials.

Thermoelectric refrigeration is at least an order of magnitude more costly to operate and has been shown to cost more than a comparable, conventional appliance by a factor of about 20.

Despite these adverse factors, Peltier devices are being used widely to cool sensitive electronic circuitry. It is virtually trouble-free and can be controlled relatively simply.

11.8.7. Hall-Effect Devices

An analysis of the Hall effect is given in Section 11.1.3. This property serves as an important means for the evaluation of the properties and quality control of semiconductors. It gives the number and sign of the carriers and the mobilities provide a measure of scattering. These parameters make it an important scientific and technological tool.

It was shown that

$$V_H = R_H IH/d \tag{11-40}$$

and that

$$R_H = \frac{1}{ne} = \mu\rho \tag{11-38}$$

Therefore, for large values of V_H, I and H should be large and n and d should be small. Many Hall devices are made very thin because of this. The result is that Joule heating becomes an important consideration and limits the amount of current which may be used. Therefore, it is necessary to examine other factors in order to optimize V_H.

Equation 11-40 may be rewritten as

$$V_H/H = R_H I/d \tag{11-40a}$$

The current can be obtained from the Joule heating as

$$Q = I^2 R = j^2 \rho (L/wd); \; I = (Qwd/\rho L)^{1/2}$$

where L, w, and d are the length, width and thickenss of the Hall specimen, respectively. This expression for I can be used along with Equation 11-38 in Equation 11-40a to obtain

$$V_H/H = R_H I/d = \mu\rho(Qwd/\rho L)^{1/2}/d$$

and may be rewritten in the form

$$V_H/H = \mu\rho^{1/2}(Q/wL)^{1/2}(w/d^{1/2})$$

This relationship is helpful in the design of devices based upon the Hall effect. Since most devices of this kind are very thin, and have much more area than volume, Q/wL provides a measure of the heat generated by the device. This factor can be optimized, along with $w/d^{1/2}$, to determine the geometry of the device, as well as I and Q. These, then, become fixed for the given application. Once this is done, $\mu\varrho^{1/2}$ becomes a figure of merit for the device. It is applicable to all d.c. Hall devices.

Some of the better semiconductor materials used in Hall devices include n-type InSb, with a figure of merit, near room temperature, of about $40 \times 10^3 \Omega^{1/2}$ cm^2/Vsec. and n-type Ge and Si each with values of about $20 \times 10^3 \Omega^{1/2}$ cm^2/Vsec. These materials are used in such devices as gaussmeters, ammeters, oscillators, switches, and wattmeters.

11.9. Semiconductor Diode Devices

The devices discussed in the following sections have been included as being representative of some of the more commonly used diodes. The intent is to demonstrate many of the more important physical mechanisms involved, rather than to provide a catalogue of devices.

An important class of semiconductor diodes is based upon single p-n junctions. The readily controlled charge carriers and electric fields in the depletion zones in the neighborhood of junctions make these devices possible (Section 11.6). The variations in the behaviors of the carriers and the internal electric fields when influenced by voltages, radiation, and other external effects determine the properties, type, and applications of the devices. Since the junctions are small, and the reactions take place near them, such devices can be, and are, made to be very small.

The graphs are intended to show functional relationships, rather than actual values because of the variations in the properties of devices made by different sources.

11.9.1. Semiconductor Diode Rectifiers

The polarity of the power supply is important in many electronic circuits. Power of the wrong sign can incapacitate many devices. Protection against this event may be obtained by the use of two oppositely oriented rectifiers in parallel between the supply and the circuit. Current of the wrong sign will be blocked by one of the rectifiers (negatively biased) and the circuit will be protected.

In a comparable way, excessive voltage (overvoltage) can be eliminated by means of two oppositely oriented rectifiers in series between the input and the circuit. These limit the input voltage to the desired value without entering into the circuit.

Two oppositely oriented rectifiers in parallel with each other, and connected in parallel with a suitable resistor across the terminals of a circuit, can provide protection against excessive current. In such an event, the rectifiers limit the excess current and allow only the desired amount of current to enter the circuit.

Rectifying diodes may be used as voltage regulators. This is based upon their behavior when forward biased. Once a.c. current starts to flow, the voltage stays within relatively narrow limits. The initial portion of the curve in Figure 11-21 can be made to be nearly zero for low voltages ($< \sim 0.6$ V) and to have relatively flat, steep slopes at higher voltages (> 0.8 V). This results from the fact that the resistance of the junction is greater than that of the bulk material. At higher voltages, the bulk resistance predominates and is virtually constant (at constant temperature) and results in nearly linear I vs. V curves with very steep slopes.

Rectifying diodes can be used in suitable circuits to remove, or "clip", the maximum or minimum voltages, or both, from a.c. supplies. Pairs of similarly oriented diodes, in appropriate circuits, may be used for half-wave rectification. Full-wave rectification may be obtained by the use of four rectifying diodes for d.c. power supplies.

11.9.2. Avalanche and Tunnel Diodes

Breakdown occurs in an avalanche diode at a given, reverse voltage (Section 11.6.1). Diodes are manufactured to have specific breakdown voltages. Very small increases beyond such voltages result in very large increases in current. Diodes of this kind are available with power ratings from 250 mW to 50 W. The currents produced by these devices are limited by their power ratings, or burn-out will occur.

One application of avalanche diodes is for the protection of meters. The diode is placed in the meter circuit so that if excessive voltage is applied, virtually all of the current will flow thorugh the diode rather than the meter circuit.

Since the avalanche breakdown voltage of a given diode is very nearly constant, it can be used as a voltage reference. Certain diodes used for this purpose have nearly equal avalanche and tunneling effects at about 5.3 V are minimiumly affected by temperature because the tunneling and avalanche effects have opposite temperature coefficients which approximately cancel each other. This phenomenon also makes them useful in voltage-stabilization circuits.

The very high doping which is required for tunneling to take place (Section 11.6.1) is responsible for the junction being in reverse, internal breakdown in the absence of an applied voltage. As a forward bias is applied it begins to counteract the internal breakdown until a maximum current is reached (Figure 11-27). The internal breakdown then diminishes with increasing voltage up to about 0.3 to 0.4 V. This portion of the curve is called the negative resistance region. Beyond this voltage range, the current-voltage curve is like that of the usual p-n rectifying junction.

The response time in Zener, or tunnel, diodes is very short. Use has been made of this property to employ these devices in relatively simple circuits for oscillators in the GHz (10^9 cps) range and for high-speed, pulse circuits ($\sim 10^{-9}$ sec). Such very short response times also make them desirable for switching devices. In addition, they are used for microwave receivers because of their negligible breakdown voltages.

Tunnel diodes also are useful for converting the output of low-voltage, high-current devices to higher voltages. One such application has been their use in conjunction with thermoelectric generators.

11.9.3. Photodetecting Junction Devices

The electron-hole pairs created by photons near a p-n junction are accelerated by the large internal field in the space-charge region, causing a current to flow in a circuit (Sections 11.6 and 11.9.6). Any photo-excited minority carriers outside of this region can diffuse toward it, depending upon their lifetimes. Once they approach the depleted region, they too, are accelerated by the internal field and augment the current. There-

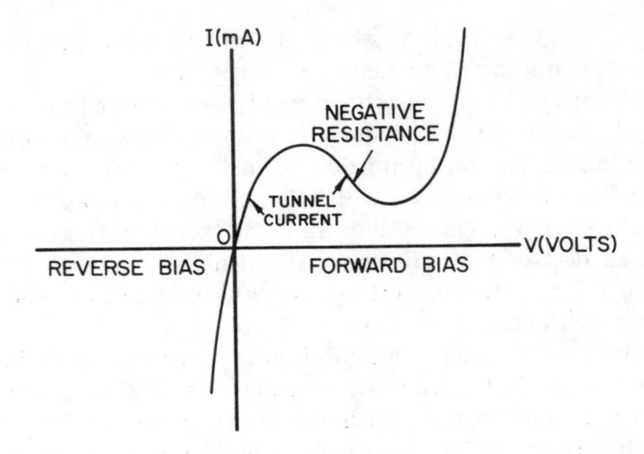

FIGURE 11-27. Current-voltage curve for a tunnel diode.

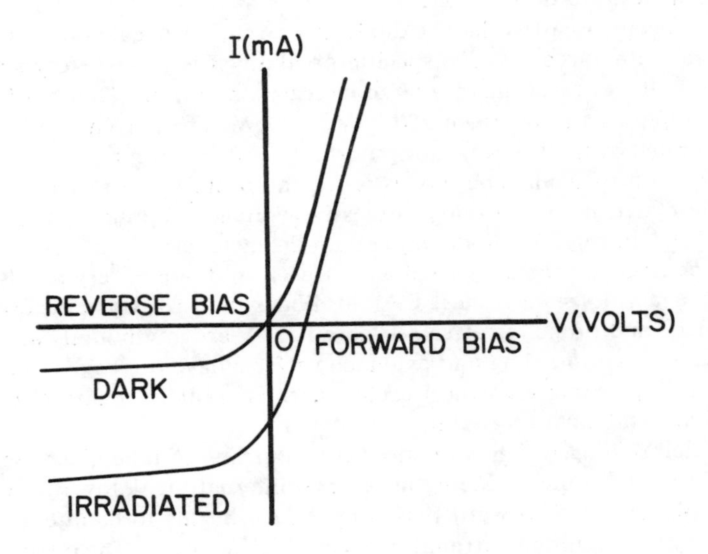

FIGURE 11-28. Characteristics of a photodetector. (After Beeforth, T. H. and Goldsmid, H. J., *Physics of Solid State Devices,* Pion Ltd., London, 1970, 102. With permission.)

fore, in addition to the depleted region, volumes of about one mean-free-path length on either side of it play an important part in these devices. The mechanism is known as the photovoltaic effect. These devices convert electromagnetic energy into electrical energy; they are known as solar cells when used to convert sunlight to electrical energy.

These devices may be used to amplify photon-induced current if a reverse bias is applied; this must be large enough to cause avalanching to occur. However, the current produced in the third and fourth quadrants of Figure 11-28 is caused by radiant energy without reverse bias. This results from current flow from the positive terminal and gives a negative current. The resulting power (W = IV) is negative, indicating that power generation, not utilization, is occurring.

Se cells are most effective for photographic purposes, since the wavelength at which their maximum response occurs is close to that of the human eye and to that of sunlight.

Radiation sources in the near-infrared are detected more effectively by Si cells. The efficiency of Si cells is about 11 to 13%, with a possible maximum of about 20%. Si cells are used in preference to Se cells because they are about 20 times more efficient. CdS cells are about half as efficient as Si cells being about 7%. However, their low cost compared to Si makes them commercially attractive (see Section 11.8.1).

11.9.4. Light-Emitting Diodes

Semiconductor diodes that emit electromagnetic radiation in the visible portion of the spectrum are called light-emitting diodes (Section 11.3). The current-voltage curves for these diodes are similar to that shown in Figure 11-21. The threshold voltages for current flow, when forward biased, will be different for different materials. For example, Ge, Si, and GaAs have threshold voltages of about 0.2, 0.6, and 1.0 V, respectively. Ge and Si emit only small amounts of infrared light, while GaAs is a source of much larger amounts of infrared radiation. GaAsP materials with thresholds of between 1.4 and 1.8 V, depending upon the P content, emit visible light in the range from red to amber.

The application of a forward bias injects electrons from the n-type into the p-type conduction band. Recombination occurs when they drop back to the valence band of the p-type material and emit a photon. However, the energy may be nonradiative and be given off as phonons. If significant amounts of trapping occur in the recombination process (Section 11.2.1), much of the energy may be absorbed as phonons. The most efficient photon-producing processes occur when direct recombination takes place. The energy of the emitted photon is determined by E_g. This is the equivalent of saying that E_g determines the wavelength of the emitted light (Section 11.3).

The efficiency of internal light generation is high, but very little of it is emitted from the device. Much of the light is absorbed internally because of the long paths and low transparencies of the materials. In addition, some of the light that reaches the surface is reflected back by the surface and is absorbed.

Light-emitting diodes find considerable application in pilot lights and in digital readout devices. These have very long lives ($\sim 2 \times 10^5$ hr). In addition, their usefulness is increased by their compatibility with transistor logic circuitry.

11.9.5. Semiconductor Diode Lasers

Electroluminescent diodes give off light as a result of the promotion of electrons from the valence band caused by the injection of electrons. The excited electrons emit a photon (spontaneous emission) and drop back to the valence band. Diodes such as these are costly and are unable to produce high light intensities. Consequently, only those applications already noted are made of these diodes (Section 11.9.4).

The best of the early semiconductor diode lasers were made from p-n junctions in GaAs. They were small, required about 2V to operate and were rugged and inexpensive. The high current densities required for their operation made it necessary that they be operated in short pulses ($\sim 10^{-6}$ sec) at ordinary temperatures. The time between pulses was about 10^{-3} sec. This combination provided adequate pumping without overheating. Continuous operation was possible only at cryogenic temperatures (77 K and lower).

The drift of electrons across a forward-biased junction brings them into the conduction band of the p-type region where they have energies of about E_g. Electroluminescence occurs at low current densities. At higher current densities (2.5 to 10×10^4 amp/cm^2), pair production increases and stimulated emission and amplification (Section 11.8.2.2) occur. The optically flat ends, perpendicular to the junction, may serve as partial mirrors and a beam of coherent light is emitted from the p-type area adjacent to and parallel with the junction. Diodes such as these are called homostructure lasers because both parts of the diode are made from the same material, doped GaAs.

The lasing region in GaAs diodes is wide because the mean-free path of an injected electron may be of the order of several microns. This is the reason for the high, critical, current density required for lasing in homostructure lasers. In addition, the wide lasing volume also results in the absorption of many photons rather than in photon amplification. Thus, the intensity of the emergent beam is diminished by the high internal losses.

These deficiencies are overcome by the use of what is known as a double heterojunction. One of these devices is shown schematically in Figure 11-29a. The central region is p-type GaAs. On either side of this are regions of n- and p-type $Al_xGa_{1-x}As$. The $Al_xGa_{1-x}As$ has a larger value of E_g than does GaAs. The increase in E_g is directly proportional to the amount of Al ions present. These layers are sandwiched between n- and p-type GaAs, respectively.

The larger values of E_g in the $Al_xGa_{x-l}As$ layers have very important functions in these heterojunctions. The junction between the n-type $Al_xGa_{l-x}As$ and p-type GaAs (Figure 11-29b) results in a barrier for the holes in the p-type GaAs, but the conduction band is unchanged. When this junction is biased to inject electrons into the p-type GaAs, the holes are prevented from enetering the n-type $Al_xGa_{l-x}As$. In the junction between the p-type GaAs and the p-type $Al_xGa_{l-x}As$ (Figure 11-29c) the valence band remains unchanged, but the width of the conduction band in the $Al_xGa_{l-x}As$ is less than that of the p-type GaAs because of the larger value of E_g. This constitutes a potential barrier which prevents injected electrons in the conduction band of the p-type GaAs from entering that of the p-type $Al_xGa_{l-x}As$. This barrier is unchanged by forward biasing. Thus, both types of carriers are contained within the central, p-type GaAs. Therefore, recombination and the resulting photons are generated in a very small volume. The light is contained within the p-type GaAs layer because of the different refractive indexes of GaAs and AlGaAs.

The constraints which result from the heterojunction prevent the electrons and the holes from diffusing out of the junction into a wider volume (curved arrows in Figure 11-29a) and the abrupt changes in the indexes of refraction contain the light within a small volume. When a forward bias is applied, electrons are reflected at the p-p junction and the holes are reflected at the p-n junction. Since both types of carriers are contained within the small volume, much lower current densities (1 to 3 × 10^3 amp/cm^2) are required for lasing. The thin, active, p-type, GaAs layer constitutes a much smaller volume than a homojunction for the confinement of the coherent, standing waves.

Even though the finished wafers containing these heterojunctions are small (about 0.5 mm × 0.1 mm × 75 to 125 μm thick), the heat dissipation necessitated by the required current densities is difficult to accomplish. This is a result of the extremely low thermal conductivites of the materials involved (Section 4.4 and Figure 4-20, Volume I). Devices with smaller active zones, but with integral heat sinks have been made. These have operated continuously at temperatures as high as 100°C. Energy conversion efficiencies of more than 10% have been reached.

Some of the more common laser applications are given in Section 11.8.2.2.

11.9.6. Parametric Diodes

Almost all of the devices discussed up to this point have been considered primarily under conditions of forward bias. Reverse bias is given primary attention here because it provides further insight into the nature of the depletion zone and explains the operation of many useful devices. These include varactors, p-i-n junctions, Schottky diodes, and IMPATT diodes.

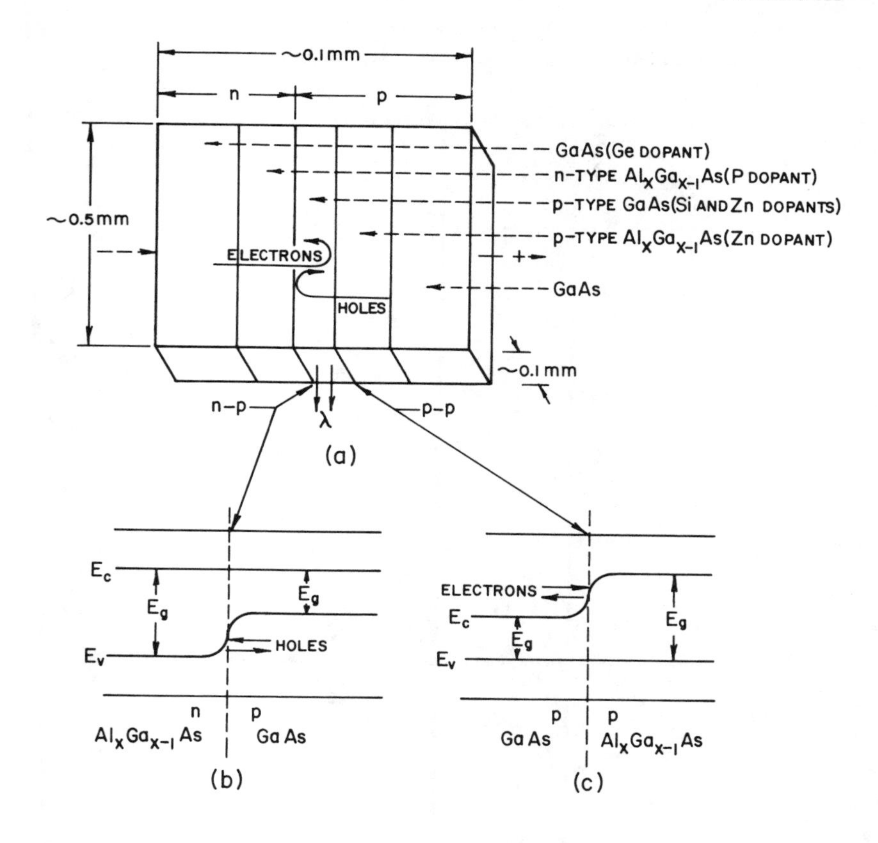

FIGURE 11-29. (a) Composite, layered, double heterojunction. Emergent arrows indicate the coherent beam. (b) n-p junction. (c) p-p junction.

11.9.6.1. Varactors

Varactors are diodes whose properties and applications are based upon the behavior of the capacitance of the depletion zone as a function of voltage. The junction, again, is considered to be of zero width (the concentration gradients at the junction are taken as being step functions). The properties of interest are shown schematically in Figure 11-30, using a one-dimensional analysis, with the x axis along the axis of the device. The internal space charge sets up electric fields in the depleted regions on either side of the junction (Section 11.6). The variation of the internal field with distance on the n side of the junction (Figure 11-30b) is given in terms of the number of ionized donors, N_d, where ε_D is the product of the dielectric and permitivity constants, as

$$\frac{d\bar{E}}{dx} = \frac{eN_d}{\epsilon_D}$$

Upon integration

$$\int_0^{\bar{E}} d\bar{E} = eN_d/\epsilon_D \int_{x_n}^{x} dx; \; \bar{E}_n = eN_d(x - x_n)/\epsilon_D$$

The value of the limit of x_n of the integral for the internal electric field may be taken as zero because the field external to the depleted zone is very small compared to that within it. At the junction, $x = 0$ and

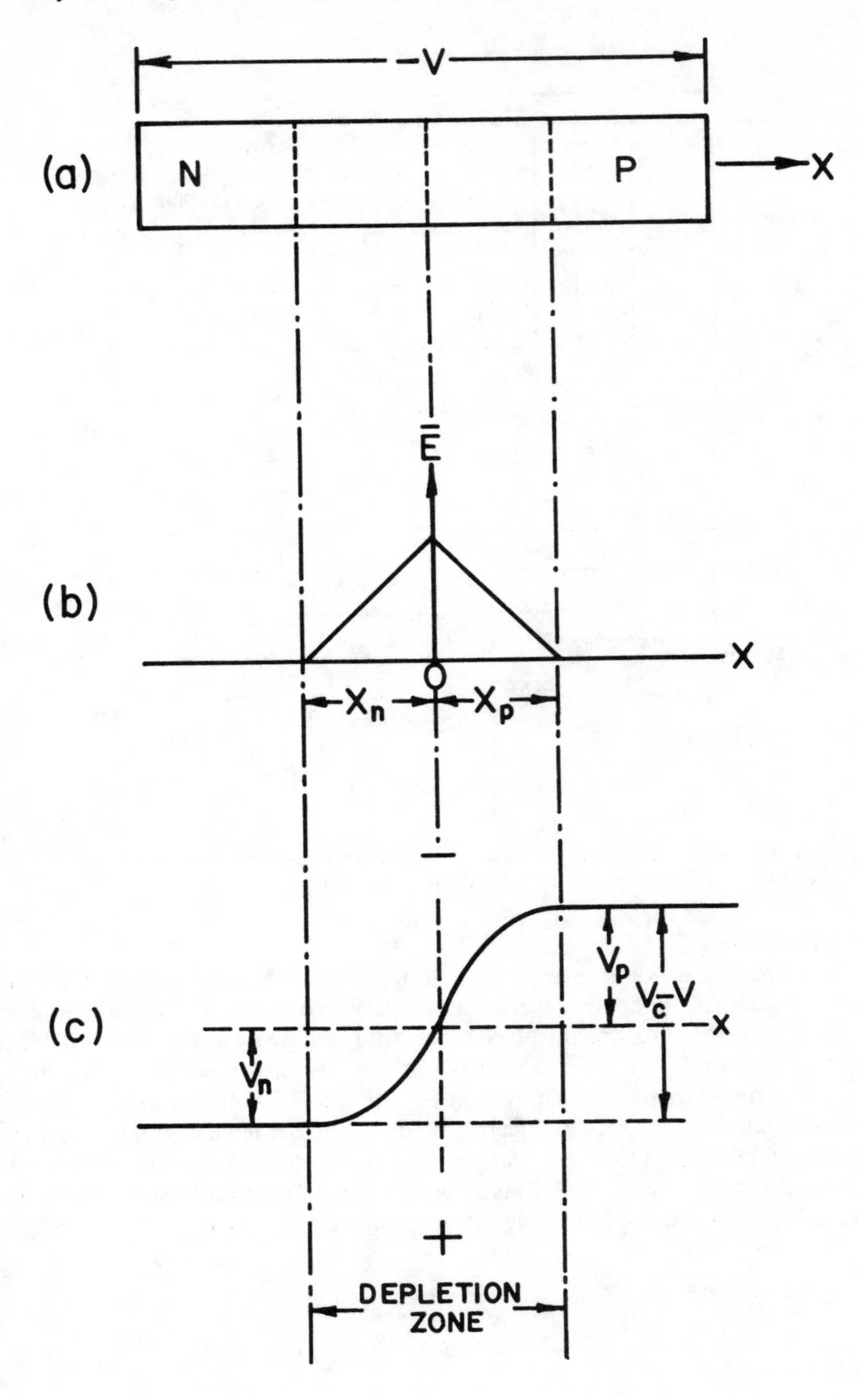

FIGURE 11-30. (a) Reverse bias diode (with depletion zone greatly enlarged for clarity). (b) Electric field. (c) Potential variation. (After Leck, J. H., *Theory of Semiconductor Junction Devices*, Pergamon Press, Elmsford, N.Y., 1967, 93. With permission.)

$$\bar{E}_n = -eN_d x_n/\epsilon_D$$

Similarly, for the internal field at the p side of the junction,

$$\bar{E}_p = eN_a x_p/\epsilon_D$$

where N_a is the number of ionized acceptors. The field must be continuous, so that, at the junction where the fields must meet, $\bar{E}_n = \bar{E}_p$ and $|N_d x_n| = N_a x_p$ (x_n is negative and $|x_n| \neq x_p$).

The potential across the region from zero to x_n is obtained from

$$\frac{dV}{dx} = -\bar{E}_n = -eN_d(x - x_n)/\epsilon_D$$

Integrating and assigning V = 0 at x = 0 gives

$$V_n = -eN_d(x^2/2 - xx_n)/\epsilon_D$$

At the edge of the depleted zone $x = x_n$, so

$$V_n = eN_d x_n^2/2\epsilon_D$$

In a similar way, for the p region,

$$V_p = -eN_a x_p^2/2\epsilon_D$$

The voltage drop outside of the depletion zone is comparatively small and may be neglected. The potential difference in the space-charge region must equal the difference between the contact potential difference and the applied voltage (Figure 11-30c). Thus, where V_c is the contact potential difference* and V is the applied voltage, and using the above equations for V_n and V_p,

$$|V_c| - V = V_n + V_p = e(N_d x_n^2 + N_a x_p^2)/2\epsilon_D$$

This may be rewritten as

$$|V_c| - V = \frac{e}{2\epsilon_D}\left[\frac{N_d^2 x_n^2}{N_d} + \frac{N_a^2 x_p^2}{N_a}\right]$$

Recalling that $|N_d x_n| = N_a x_p$ and substituting this in the above equation gives

$$|V_c| - V = \frac{e}{2\epsilon_D}\left[\frac{N_a^2 x_p^2}{N_d} + \frac{N_a^2 x_p^2}{N_a}\right]$$

This expression is rearranged to read

$$x_p^2 = \frac{2\epsilon_D}{e}(|V_c| - V)\left[\frac{N_a N_d}{N_a + N_d}\right]\frac{1}{N_a^2}$$

* V_c can be negative when the density of intrinsic carriers $N_i^2 \ll N_a N_d$ and $N_a > N_d$. Using $N_i^2/N_a N_d \simeq \exp(eV_c/k_K T)$, e is positive and V_c is negative.

$$x_p^2 = \frac{2\epsilon_D}{e}(|V_c|-V)\left[\frac{N_d}{N_a(N_a+N_d)}\right] \tag{11-76}$$

Where the conductivity in the n-type material is very much greater than that in the p-type, $N_d \gg N_a$ and

$$x_p^2 \simeq \frac{2\epsilon_D}{e}(|V_c|-V)(1/N_a) \tag{11-77}$$

or

$$x_p \simeq [2\epsilon_D(|V_c|-V)/eN_a]^{1/2} \simeq d \tag{11-77a}$$

Under these conditions, $x_p \gg x_n$ and the width of the space-charge region, d, may be approximated as $d \simeq x_p$. Therefore, the width of the region may be approximated as varying as $V^{1/2}$ when reversed biased. It also is important to note that, in this approximation, the extent of d is only a function of the lower conductivity material.

The space-charge region is virtually devoid of carriers. This region, then, is an excellent insulator which is sandwiched between two relatively high conductivity materials, when compared to the space-charge region. This configuration constitutes a capacitor. The capacitance is given by

$$C = \frac{A\epsilon_D}{d} \text{ (farad)}$$

where A is the area (m^2) of a parallel-plate capacitor. Now, using Equation 11-77a, the capacitance of the depleted zone is

$$C = A\epsilon_D[eN_a/2\epsilon_D(|V_c|-V)]^{1/2}$$

or

$$C = A[e\epsilon_D N_a/2(|V_c|-V)]^{1/2} \tag{11-78}$$

Since d increases with V, C decreases with V. This makes the capacitance of the junction a function of $V^{-1/2}$, hence, the names "parametric diode" or "varactor" (Figure 11-31a).

Some applications of these devices include tuning, switching, logic circuits, and parametric amplification where high speed and stability are needed. One very common use is for automatic frequency control for radio receivers.

Where the impurity concentrations on either side of the junction are "uniformly graded" (nearly linear concentration gradients which approach zero at the center of the junction), $C \propto V^{-1/3}$. Devices with this slower dC/dV behavior frequently are more useful than those with the faster rates of change shown by devices with abrupt concentration gradients.

11.9.6.2. P-I-N Junctions

The insertion of a thin layer of intrinsic semiconductor between the p- and n-type materials further enhances the capacitance-voltage effect. Devices of this kind are known as p-i-n diodes.

The operation of a p-i-n device is similar to that of a p-n junction in that holes cross

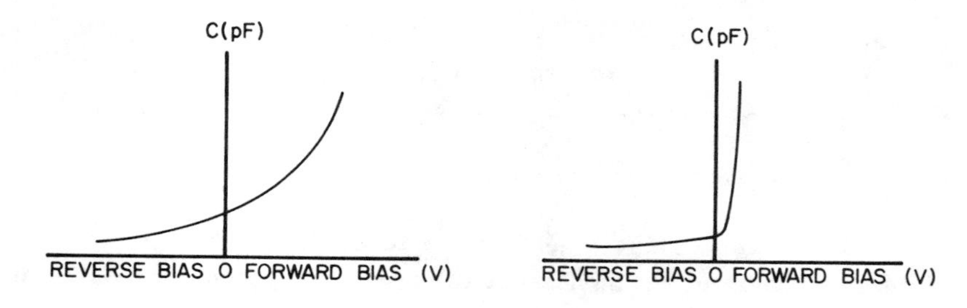

FIGURE 11-31. (a) Characteristics of a varactor, (b) Characteristics of a p-i-n device. (After Beeforth, T. H. and Goldsmid, H. J., *Physics of Solid State Devices,* Pion Ltd., London, 1970, 48 and 52. With permission.)

the i-zone to the n-type material and electrons cross in the opposite way. The carrier densities in the i-zone are very small compared to those of the other regions. Thus, the full extent of the i-region serves as a very wide depletion region and a large increase in the charge densities of the carriers occurs on the p- and n-sides of the interface. These factors cause a practically uniform electric field in the i-region. Since the intrinsic region in the p-i-n device is similar to, but larger than, the space-charge region in a varactor, their capacitance responses to external voltages are similar but smaller. However, since the applied voltages result in much smaller internal fields, a much larger voltage is required for breakdown. It is for this reason that p-i-n diodes are used where high voltages are involved (Figure 11-31b). The smaller capacitance of the p-i-n compared to the p-n device permits faster response times, and output frequency can be increased.

Some typical applications of p-i-n devices include ratio-frequency applications and switching and limiting devices. When used as photodetecting, junction devices, p-i-n diodes are more sensitive than p-n junctions because the higher fields which can be applied greatly increase pair production (see Section 11.9.3). The advantages of p-i-n devices are that they also can operate with power supplies with smaller voltages and are stable, small, inexpensive, and rugged.

11.9.6.3. Schottky Diodes

The Schottky diode consists of a semiconductor separated from a metal by a thin, oxide film (see Section 11.5). The oxide film must be kept small in order to permit the carriers to tunnel across it (Section 11.6.1). The very thin oxide films normally present upon many semiconductor materials frequently are sufficient for this purpose. The device is completed by the deposition of a metal layer upon the oxide film.

When the work function of the metal is larger than that of the semi-conductor, electrons will flow from the semiconductor to the metal (Figure 11-18). The excess negative charge builds up at the surface of the metal until it is sufficient to prevent further electron flow. At equilibrium, in the absence of an external field, the net flow of carriers is the same in each direction. Forward biasing decreases the barrier; reverse biasing increases the barrier (Figure 11-19), and the current of electrons from the semiconductor to the metal will, accordingly, increase or decrease. Where the dopant concentration is high, the depletion zone will be very narrow and additional electrons may tunnel across the barrier. Holes also may tunnel if the value of $(W_M - W_S) \simeq Eg$ (Figure 11-18).

The Schottky potential barrier, V_S, may be obtained from Equation 11-77 as

$$|V_c| - V \simeq \frac{eN_a x_p^2}{2\epsilon_D} = V_S \qquad (11\text{-}79)$$

The capacitance of the space-charge region is obtained by again letting $N_d \gg N_a$ and Equation 11-78 is obtained, for this case, from Equation 11-77a. Under reverse bias Equation 11-72 may be written more generally as

$$j = j_S [\exp (eV/k_BT) - 1] \qquad (11\text{-}72b)$$

for the mechanisms involved here, where j_s is the Schottky current density.

The width of the barrier, in the absence of an external potential, may be obtained from Equation 11-79 as

$$V_S \simeq \frac{eN_a x_B{}^2}{2\epsilon_D} \; ; \; x_B \simeq (2\epsilon_D |V_c|/eN_a)^{1/2} \qquad (11\text{-}79a)$$

in which x_B is the total barrier width.

Other mechanisms contribute to j_s, depending upon the materials. When the dopant concentration is high, x_B will be small. Then, it is possible that additional contributions may be made to j_s such as tunneling of holes and electrons, field-induced tunneling, and field emission. The latter contribution occurs when high electric fields lower the barrier significantly more than "normal" fields and more high-energy ("hot") electrons enter into the conduction process.

Schottky diodes now are made from Si or GaAs. They are used for pulse-shaping and switching and limiting functions because of the very short lifetimes of the carriers. The Schottky diode has an advantage over the ordinary diode because its current-voltage behavior in the reverse-bias mode is similar to that shown in Figure 11-21 for general p-n junctions, but gives much larger currents at saturation. The tunnel diode, in the same mode gives smaller currents at relatively small voltages. This may be expressed for ordinary, p-n Ge devices as

$$j_S \sim j_o \times 10^6$$

In other words, the Schottky diode saturation current density is about 10^6 times larger than that of an ordinary Ge device. Another important characteristic is its short, turn-off times noted above in terms of the very short carrier lifetimes. The mean-free-paths of the "normal" and "hot" electrons are very short.

11.9.6.4. IMPATT Diodes

An IMPATT diode (impact, avalanche, transit-time diode) consists of a d.c., reverse-bias, p-n junction in which the reverse voltage is maintained at a level slightly less than that required for breakdown. It is placed in the circuit in such a way that a given time-lapse occurs before the circuit reacts to avalanching.

When an alternating current is transmitted through the reverse-biased device (Figure 11-32a), the number of electrons in the depletion zone will decay as a result of avalanching and the density of holes will increase. These pass out of the depletion zone and an additional lag occurs in the circuit because of the lower mobility of the holes. The remaining current pulse is shown in Figure 11-32c. The wave-shape is changed and the resulting current is 180° out of phase with the original.

The use of these devices in tuned circuits results in continuous oscillation. P-i-n diodes also are used for this type of application. The larger voltages possible with a p-i-n device, and the high degree of carrier depletion upon reverse bias, permit their use at higher power levels. Both types of devices are very useful in the generation of microwaves. They are competitive with klystron and magnetron tubes. They have the additional advantage of compatibility with solid-state circuitry.

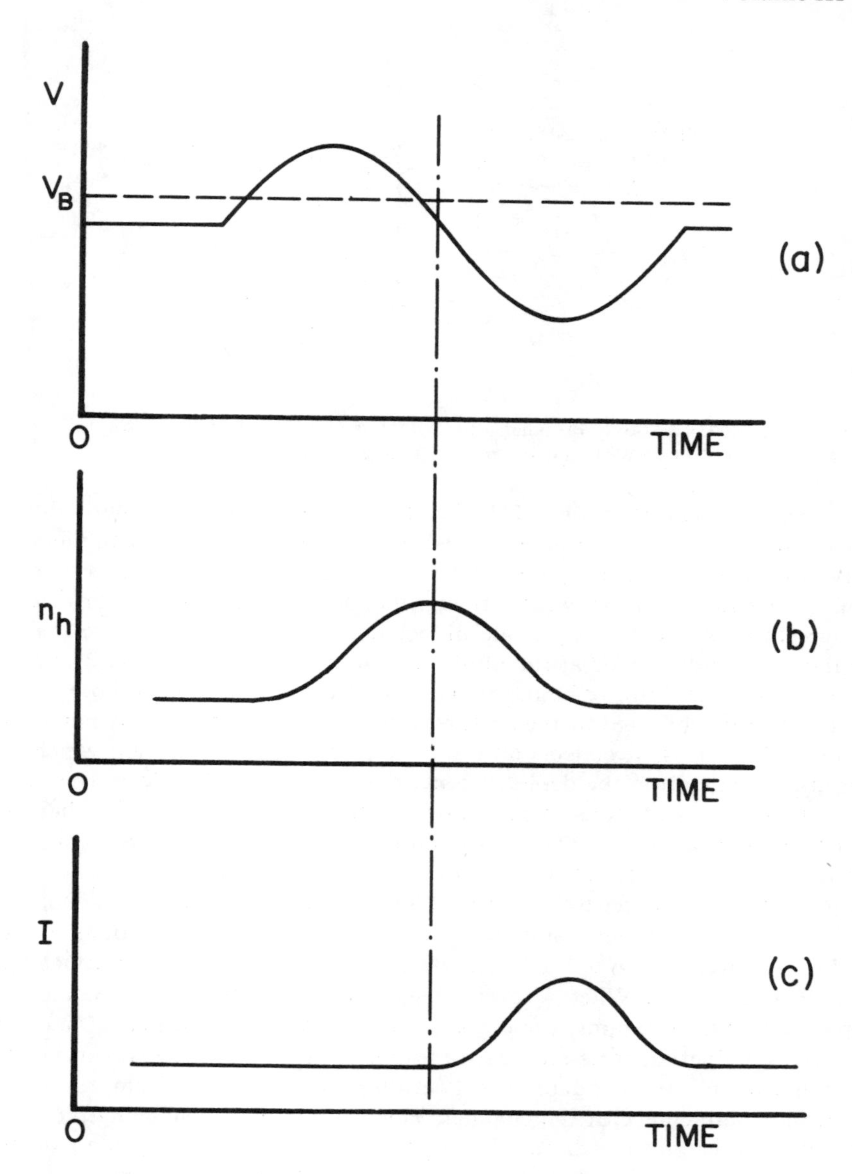

FIGURE 11-32. (a) Reverse bias, V_B, and a single cycle of AC. (b) Density of holes in depletion layer after avalanche. (c) Current delay. (Adapted from Beeforth, T. H. and Goldsmid, H. J., *Physics of Solid State Devices,* Pion Ltd., London, 1970, 55. With permission.)

11.10. TRANSISTORS

Transistors may be classified into two major kinds: bipolar-junction transistors (BJT) and field-effect transistors (JFET). Both types are named for their operating mechanisms. BJTs are based upon the contributions of both holes and electrons, thereby deriving their name. JFETs are based upon the flow of a single type of carrier, either electrons or holes (but not both), in which the resistance or the current of the device may be controlled by an electric field. JFETs, thus, are unipolar devices.

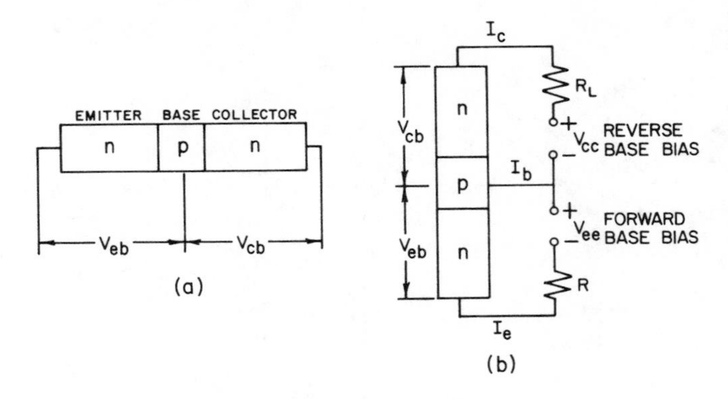

FIGURE 11-33. (a) Schematic diagram of a bipolar junction transistor (the base width is gratly enlarged). (b) Circuit.

The general configuration of a BJT consists of two regions of similarly doped material separated by a thin, oppositely doped region. This amounts to two diodes which are very close to each other. These diodes have the same properties, as described in Sections 11.6 and 11.9.1. However, their small separation makes a very significant difference. In a reverse-biased, single diode, the minority carriers have their origin within the semiconductor or at the ohmic contact. However, in the BJT, the second diode can be a large source of minority carriers because it is so very close to the first diode. The current collected by the first, reverse-bias diode is essentially not a function of the amount of bias; it collects all minority carriers inside a region which extends one mean-free-path from the depleted zone, regardless of their origin. Therefore, the number of carriers, and, consequently, the current collected by the first diode can be controlled by the voltage across the second diode. The control of the voltage of the second diode is the basis for the application of these transistors.

The junction, field-effect transistor (JFET) also may consist of n-p-n or p-n-p configurations. The central material is called the channel. Ohmic connections connect the channel to a source and to a drain. The outer materials are connected to a single terminal known as a gate. When a reverse bias is applied between the channel and the two oppositely doped regions, the depleted zone moves into the central material and the effective conduction cross-section of the channel is reduced. Thus, varying the bias between the channel and the gate, or between the source and the drain, results in variations in the conductance of the channel. This is the same as changing the resistance between the source and the drain. Virtually no current is drawn by the gate because the junction is reverse-biased. If the voltage across the channel (between source and drain) is small compared to that required for the effective reduction of the channel, the resistance of the channel will be independent of the drain current. When this is the case, the JFET becomes a linear resistor; its resistance may be varied with the voltage. However, when the voltage across the channel effectively reduces the width of the channel at the drain side, the drain current is virtually independent of the voltage between the source and the drain. When this is the case, the JFET becomes a current source in which the current is practically independent of the voltage between the source and the drain. When this is the case, the JFET becomes a current source in which the current may be varied by the voltage between the gate and the source.

Simpler, single-junction, JFET devices may be made from a single p-n junction. The source and the drain are connected to either side of one member of the junction. This half of the device contains the channel. The gate is connected to the other portion of the device. The operation of this device is the same as that described above. This simple device will be used for purposes of explanation in Section 11.10.2.

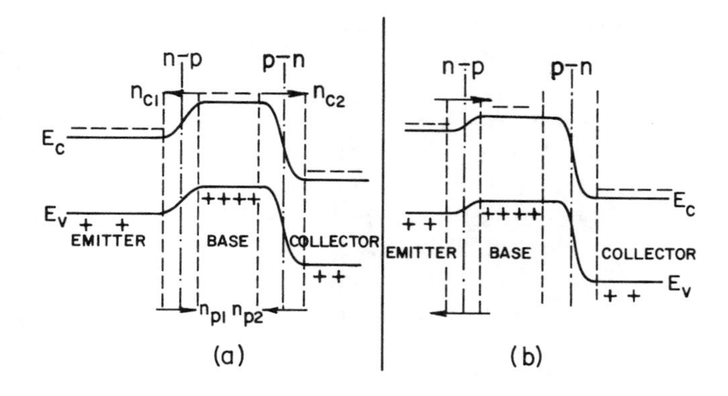

FIGURE 11-34. (a) Band structure for reverse bias on both junctions. (b) Forward bias on both junctions.

The operation of both types of devices is described in the following sections. Both are readily adaptable to microminiaturization.

As in the previous sections, the graphs are intended to show functional relationships rather than actual values because of the variations in the properties of devices made by different sources.

11.10.1. Bipolar Junction Devices

BJTs are composite devices consisting of three layers. The layers may be n-p-n or p-n-p, the central material in each case being very thin. Both configurations operate in the same way, but the polarities of the charges and potentials will be opposite in each case. The n-p-n device is discussed here. In any event, the BJT consists of two p-n junctions in close proximity as shown in Figure 11-33a.

When V_{eb} and V_{cb} are made negative, both of the junctions are reverse biased; only the very small, reverse-bias currents will flow (Figure 11-21a). This current is the sum of the two reverse currents. The band structure adjacent to the two junctions is shown schematically in Figure 11-34a. This reverse current must be very small because the density of carriers flowing between the emitter and the base is the sum $n_{c1} + n_{p1}$ and the density flowing from the base to the collector is $N_{c2} + n_{p2}$. Each of the components of these sums virtually cancel each other. When this condition is present, the transistor is "off" and it behaves as though the circuit was open. This is the first mode of operation and is used for cut-off circuitry applications.

When a forward bias is imposed upon the base (Figure 11-34b) the collector-base (n-p) junction again will be reverse bias. Using the Boltzmann statistics, the number of electrons (minority carriers) per unit volume adjacent to the depletion zone at the n-p junction (emitter-base junction) is

$$n_{e(n,p)} = n_e \exp(eV_{eb}/k_BT)$$

And, in the same way, the number of minority carriers per unit volume on the base side of the depletion zone at the p-n junction (base-collector junction) is

$$n_{e(p,n)} = n_e \exp(eV_{cb}/k_BT)$$

It can be seen from the difference in the barriers at the two junctions that $n_{e(n,p)} > n_{e(p,n)}$. And, since the base contains a large number of majority carriers, virtually no internal field is present in this region. Thus, no opposition exists which would prevent the diffusion of the electrons across the base, to the collector, at a fairly uniform rate. This leads to a linear excess charge distribution where $n_{e(n,p)} \rightarrow n_{e(p,n)} = 0$ at the base-collector interface.

The emitter dpoant concentration is made significantly higher than that of the base so that forward bias across the n-p junction results in virtually all carriers originating from the emitter entering the base. (This also is called carrier injection into the base.) This almost entirely precludes carrier injection from the base to the emitter. The carrier migration from the emitter to the collector is optimized by keeping the base very thin. This minimizes recombination and allows almost all of the carriers from the emitter to enter the collector. In other words, almost all of the current will enter the collector. In actual devices, the fraction of the current entering the collector from the emitter, α, is at least 0.95 and may reach 0.98 to 0.99+. However, if the temperature becomes too high, the number of carriers becomes excessive and "thermal runaway" takes place.

The very small amount of recombination that does take place is important since it causes a small, external flow of current into the base. A small number of holes must enter via the base connection. This maintains the density of the majority carriers in the base very close to that of equilibrium. Therefore, this current must equal the rate of recombination of the electrons and holes in the p-type base.

The current entering a device must equal the currents leaving it (Kirchoff's current law: $\Sigma I = 0$). In the present case, this is expressed as

$$I_e = I_b + I_c \qquad (11\text{-}80)$$

where the subscripts e, b, and c denote emitter, base, and collector, respectively. The base current needed to compensate for recombination in the base is

$$I_b = (1 - \alpha)\, I_e \qquad (11\text{-}81)$$

I_b is the input current and the output current is I_c; the ratio of these is the gain. This is given by

$$I_c/I_b = \beta \qquad (11\text{-}82)$$

This ratio, β, is increased by highly doped emitter material and very thin base zones. It is reasonably constant over a wide range of I_c, but it decreases at low currents because of recombination in the depletion region of the emitter-base junction.

These relationships give insight into the characteristics of BJT devices. Equation 11-80 is rearranged and divided through by I_b to give

$$\frac{I_c}{I_b} = \frac{I_e}{I_b} - 1$$

And, by means of Equation 11-82, this becomes

$$\beta = \frac{I_e}{I_b} - 1; \ \beta + 1 = \frac{I_e}{I_b}$$

Multiplying both sides of this equation by I_c gives

$$(\beta + 1)I_c = \frac{I_e}{I_b}\, I_c = \beta I_e$$

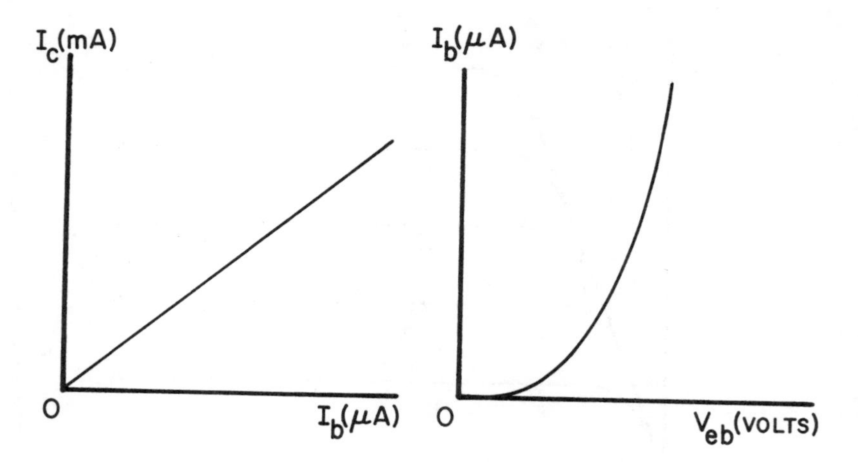

FIGURE 11-35. (a) Base and collector current relationship. (b) Base current-voltage behavior. (After Beeforth, T. H. and Goldsmid, H. J., *Physics of Solid State Devices*, Pion Ltd., London, 1970, 65. With permission.)

This is reexpressed as

$$I_c = I_e\beta/(\beta + 1) \tag{11-83}$$

Note was made previously of the fraction of the current entering the collector from the emitter. This is, using Equation 11-83,

$$I_c = \alpha I_e;\ \alpha = \beta/(\beta + 1) \tag{11-84}$$

Since $\alpha \rightarrow 1$, $I_c \rightarrow I_e$ and, from Equation 11-80

$$I_c >> I_b \tag{11-85}$$

Because $I_c/I_b = \beta$ and $I_c \gg I_b$, β is the d.c. current amplification; β may range between 20 and 500, depending upon the device. This, obviously, is useful as a means of amplification.

A transistor operated in this way shows linear behavior (Equation 11-82 and Figure 11-35a). Since the number of carriers crossing from the emitter to the base varies as $\exp(eV_{eb}/k_BT)$, the base current, I_b, also is an exponential function of V_{eb} as shown in Figure 11-35b. Many circuits make use of this exponential behavior of the base current. This is an active operating mode and constitutes a second manner of operation.

In this active mode, I_b is small and the approximation may be made that the emitter current will equal the current between the emitter and the collector, I_{ec}, and that this current will vary as the voltage between the emitter and collector, V_{ec}; $I_e \simeq I_{ec} \propto V_{ec}$. Then, using Equation 11-80,

$$I_c \simeq V_{ec} - I_b \tag{11-86}$$

This is plotted in Figure 11-36. The initial linear response corresponds to that shown in Figure 11-35a and Equation 11-82. But, when V_{ec} causes reverse bias at the collector junction, it has no further influence upon I_c, which remains essentially constant. I_b, then, becomes the controlling factor. This is known as saturation.

When V_{eb} is increased, I_c will increase and the voltage between the emitter and the collector will decrease Figure 11-33b):

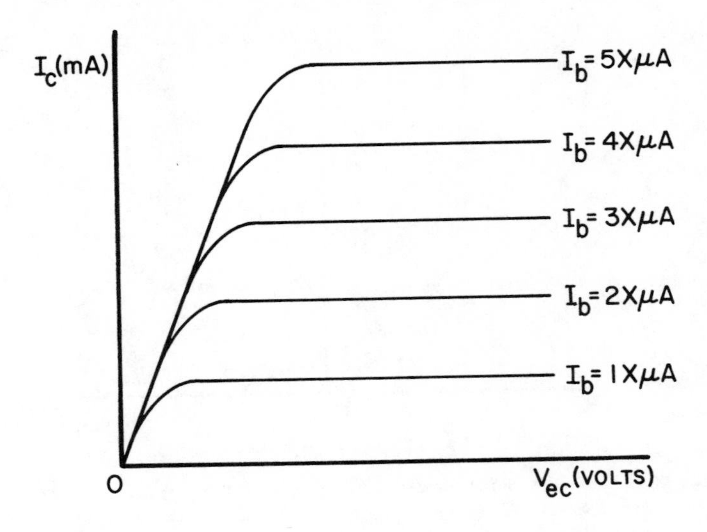

FIGURE 11-36. Response of a BJT to small signals. The ranges in which I_c remains constant as a function of V_{ec} are in the active region.

$$V_{ec} = V_{cc} - I_cR_L \tag{11-87}$$

Continued increase in V_{eb} will cause the junction to reach a condition in which V_{ec} = V_{eb} and the reverse bias vanishes. When this is the case, V_{eb} = 0 and the collector current is

$$I_c = (V_{cc} - V_{eb})/R_L$$

However, since $V_{eb} \ll V_{cc}$,

$$I_c \simeq V_{cc}/R_L \tag{11-88}$$

Again, using Equation 11-82 with Equation 11-88, the base current is

$$I_b = I_c/\beta = V_{cc}/\beta R_L \tag{11-89}$$

An additional increase in I_b in excess of that given by Equation 11-88 causes the n-p and p-n junctions to become forward biased. This causes the injection of carriers into the base from the collector and results in a current, I_{cb}. Carriers also diffuse from the emitter to the collector, giving rise to I_{ec}. The net collector current is given by

$$I_c = I_{ec} - I_{cb}$$

I_c cannot be reversed even if additional carriers are injected into the base from the collector, since the collector is positive and is connected to the positive side of V_{cc}. The value of $n_{e(n,p)}$ will decrease linearly from the n-p junction to the p-n junction by diffusion. However, because of the positive potential ($V_{eb} > V_{ec}$), $n_{e(n,p)} \rightarrow n_{e(p,n)} > 0$. This means that $V_{eb} > V_{bc} > 0$. V_{ec}, as previously noted, is small and may be neglected in comparison with V_{cc} so that it may be omitted in Equation 11-86. Thus, V_{eb} accounts for the uniform increments caused by I_b in the saturation mode (Figure 11-36). This is

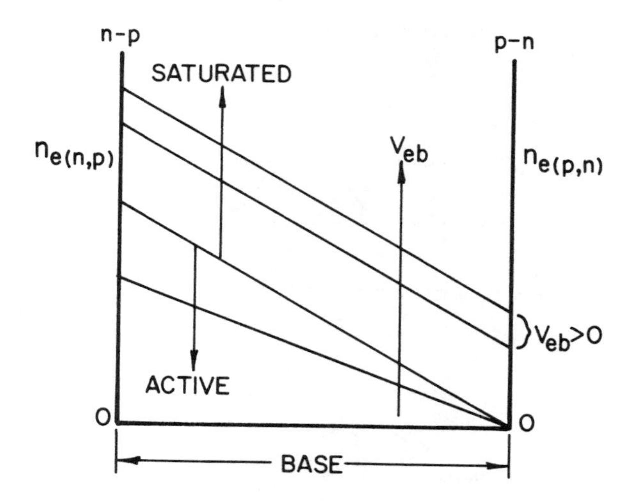

FIGURE 11-37. Differences in minority carrier concentrations between active and saturated conditions in a BJT as affected by V_{eb}. (After Beeforth, T. H. and Goldsmid, H. J., *Physics of Solid State Devices,* Pion Ltd., London, 1970 66. With permission.)

in contrast to the active mode in which the minority carrier density in the base is $n_{e(n,p)} \rightarrow n_{e(p,n)} = 0$ at the p-n junction. The influence of V_{eb} upon the operation of these devices is shown in Figure 11-37.

Since the collector current is saturated for the condition given by Equation 11-86, the effect of increasing V_{eb} is that of increasing I_b, and, therefore, increasing the minority carrier density within the base. The rate of diffusion of the carriers is essentially the same as in the unsaturated case; the main difference is that more carriers are present. This accounts for the shifts in the densities of carriers with increasing V_{eb}, the parallel slopes in the saturation region of Figure 11-37, and the uniform increments of I_c induced by constant increments of I_b in Figure 11-36.

In summary, when $V_{eb} \leqslant 0$, $I_c = 0$. When $V_{eb} > 0$, but insufficient to cause saturation, the active mode prevails and linear behavior takes place up to the limiting condition given by Equations 11-88 and 11-89. Beyond this limiting condition, the transistor is saturated, and I_c becomes constant vs. V_{ec} and is a function of V_{eb} and I_b Figures 11-36 and 11-37).

The control of V_{eb} also enables the use of the transistor as a switch. BJTs applied for this purpose may be used in the MHz to GHz range of frequencies.

11.10.2. Junction Field-Effect Transistors

Transistors based upon the field effect (FET) are very important in electronic technology, particularly in integrated circuits. The characteristic of principle interest is the depletion zone (Section 11.9.6). A depletion layer will exist in the semiconductor material adjacent to a p-n junction in the absence of an applied voltage (Section 11.6).

A single-junction FET, which now is used rarely, will be employed to explain the mechanisms involved. This is shown in Figure 11-38.

When the n-type channel material is forward biased (the p-type gate is negatively biased), V_{GS}, the extent of the depletion zone will increase uniformly in its extent into the n-type material (Equation 11-77a and Figure 11-38a). This will continue until the depletion zone extends entirely across the channel. This occurs when the gate reverse

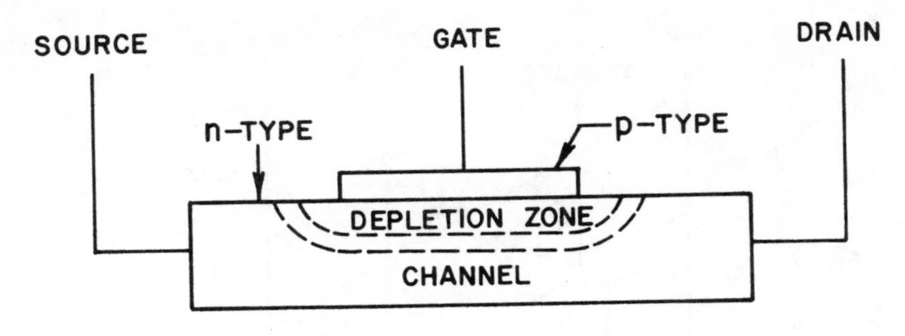

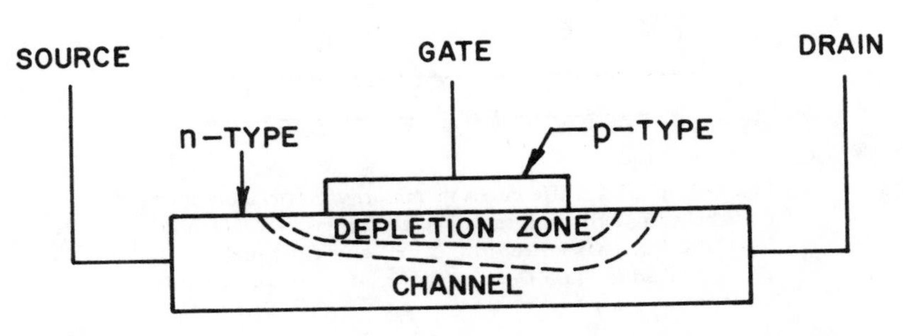

FIGURE 11-38. Schematic diagram of a JFET device with a single junction. (a) Increasing depletion zone with increasing V_{GS}. (b) Effect of increasing V_{SD} upon the profile of the depletion zone, with V_{GS} constant. (After Beeforth, T. H. and Goldsmid, H. J., *Physics of Solid State Devices,* Pion Ltd., London, 1970, 74. With permission.)

bias reaches a value V_P; the channel now is "pinched off". Using the relationship found for a Schottky barrier, Equation 11-79a may be employed to determine V_P, neglecting the contact potential, V_c, to be

$$V_P = eNx_B^2/2\epsilon_D = eNW^2/2\epsilon_D \quad (11\text{-}79b)$$

In this case, N is the number of ionized donors of a single type in the channel and $x_B = W$ is the total width of the channel material. Since the channel may be in either p-type or n-type material, N, accordingly, will be negative or positive and V_P will be positive for channels in p-type materials and negative for channels in n-type materials.

Thus, in a way analogous to the behavior of the Schottky barrier, variations in the gate-source potential, V_{GS}, change the effective cross-section of the channel. In turn, this changes the conductance (resistance) of the channel. This introduces another factor which also affects the size of the channel: the current flowing along the channel (drain current, or I_{SD}) sets up another potential difference, V_{SD}, which increases along the channel from the source to the drain. This is in addition to the V_{GS} already present. The result is that the depleted zone becomes asymmetric (Figure 11-38b) and has its greatest extent where V_{SD} is largest. This, in effect, tapers the channel; it resembles a venturi tube. The JFET acts like a resistor under these conditions. The voltage range in which this behavior occurs is known as the ohmic region.

In the operation of this device, V_{SG} may be insufficient to cause pinch-off by itself, but at a given value of V_{SD} pinch-off may occur. The critical value of V_{SD} will be denoted by $V_{P(SD)}$. This condition is expressed by

$$V_{P(SD)} = V_{GS} - V_P \quad (11\text{-}90)$$

In an n-type channel, V_{GS} and V_P are negative and $V_{P(SD)}$ is positive.

The behavior of the nondepleted, or active, region of the channel will be examined by a method based upon that of Schockley (1952). The channel is considered to be asymmetric. The potential in the channel at any point x is given by V_{SC}, with x = 0 at the source end of the n-type channel material. If the potential across the gate-channel junction is V_{GC}, then

$$V_{GC} = V_{GS} - V_{SC} \tag{11-91}$$

Let the width of the n-type channel material be W. The width of the depletion zone, based upon Equation 11-79a, is

$$x_B = [2\epsilon_D(-V_{GC})/eN]^{1/2} \tag{11-92}$$

So the width of the channel is W-x_B at any point x. This enables the calculation of the potential drop along an element of length, dx, in the channel as

$$dV_{SC} = \frac{I_{SD}\rho dx}{W - [2\epsilon_D(-V_{GC})/eN]^{1/2}} \tag{11-93}$$

in which I_{SD} is the current flowing between the source and the drain and ϱ is the resistivity of the channel. Equation 11-93 may be rearranged, and substitution made for V_{GS} using Equation 11-91, to read

$$I_{SD}\rho \frac{dx}{dV_{SC}} = W - [2\epsilon_D(V_{SC} - V_{GS})/eN]^{1/2} \tag{11-94}$$

Using Equation 11-79b, and recalling that V_P is negative for n-type material,

$$V_P = -eNW^2/2\epsilon_D; \quad -\frac{eN}{2\epsilon_D} = \frac{V_P}{W^2} \tag{11-79c}$$

Equation 11-79c is substituted into Equation 11-94 to obtain

$$I_{SD}\rho \frac{dx}{dV_{SC}} = W - [W^2(V_{GS} - V_{SC})/V_P]^{1/2}$$

or

$$I_{SD}\rho dx = W\,[1 - (V_{GS} - V_{SC})/V_P]^{1/2}\,dV_{SC} \tag{11-95}$$

Equation 11-95 is integrated from x = 0 to x = L, the length of the channel, to give

$$I_{SD}\rho L = W\left\{V_{SC} + \frac{2}{3}V_P\left[\frac{V_{GS} - V_{SC}}{V_P}\right]^{3/2}\right\} \tag{11-96}$$

The channel potential varies from V_{SC} = 0 at x = 0 and V_{SC} = V_{SD} at x = L. These limits result in

$$I_{SD}\rho L = W\left\{V_{SD} + \frac{2}{3}V_P\left[\frac{V_{GS} - V_{SD}}{V_P}\right]^{3/2} - 0 - \frac{2}{3}V_P\left[\frac{V_{GS} - O}{V_P}\right]^{3/2}\right\}$$

or

$$I_{SD} = \frac{W}{\rho L}\left\{V_{SD} + \frac{2}{3}V_P\left[\frac{V_{GS} - V_{SD}}{V_P}\right]^{3/2} - \frac{2}{3}V_P\left[\frac{V_{GS}}{V_p}\right]^{3/2}\right\}$$

and, upon factoring,

$$I_{SD} = \frac{WV_P}{3\rho L}\left\{\frac{3V_{SD}}{V_P} + 2\left[\frac{V_{GS} - V_{SD}}{V_P}\right]^{3/2} - 2\left[\frac{V_{GS}}{V_P}\right]^{3/2}\right\} \tag{11-97}$$

Neglecting the relatively small error introduced by the omission of V_c in Equation 11-79b, this analysis holds reasonably well. However, in integrated circuitry, where planar techniques are used to make functional blocks rather than individual components, the channels are made by diffusion. The concentration profile in any such diffusion process is exponential, with a negative slope. Therefore, even though the channels may be small, the dopant ion concentrations are not uniform. An approximation for the behavior in this case (Equation 11-98) is given below.

When V_{SD} approaches $V_{P(SD)}$, very high fields are produced because the channel becomes extremely narrow. A point is reached beyond which I_{SD} would decrease. This decrease would decrease V_{SD} simultaneously and cause the channel to become wider. So, when V_{SD} becomes greater than $V_{P(SD)}$, the width of the channel becomes nearly constant and the device is said to be saturated. After saturation occurs, I_{SD} remains essentially constant (Figure 11-39a). The distribution of the potentials within the channel also remain essentially constant. The saturation region is limited by breakdown which occurs when the internal fields become sufficiently large.

The saturation current may be calculated by means of Equation 11-97 by setting $V_{SD} = V_{P(SD)}$. However, for the reasons given, this is not entirely satisfactory. I_{SD} may be calculated from the relationship

$$I_{SD} = I_{SDS}(1 - V_{GS}/V_P)^2 \tag{11-98}$$

in the region $V_{SD} \geqslant 2V_{P(SD)}$, where I_{SDS} is the maximum saturation current for $V_{GS} = 0$.

The gate reacts only when the junction is in reverse bias. Thus, values of I_{GS} are in the 10^{-9} amp range. V_{GS} normally is less than 10V giving a resistance of about 10^9 Ω. This high resistance vanishes if forward biasing occurs. Therefore, V_{GS} and V_{SD} must be opposite in sign.

I_{SD} primarily is a function of V_{GS} because I_{GS} is negligible, as noted above. The amplification of a JFET is found, from Equation 11-98 as follows:

$$G = \frac{dI_{SD}}{dV_{GS}} = \frac{2I_{SDS}}{V_P}(1 - V_{GS}/V_P) \tag{11-99}$$

(The amplification, G, actually is the mutual conductance.) Solving Equation 11-98 for $(1 - V_{GS}/V_P)$, and substituting this into Equation 11-99 gives

$$G = \frac{2I_{SDS}}{V_P}\left[\frac{I_{SD}}{I_{SDS}}\right]^{1/2} = \frac{2I_{SDS}^{1/2}}{V_p} I_{SD}^{1/2} \tag{11-100}$$

G varies as $I^{1/2}{}_{SD}$, since I_{SDS} and V_P are constants.

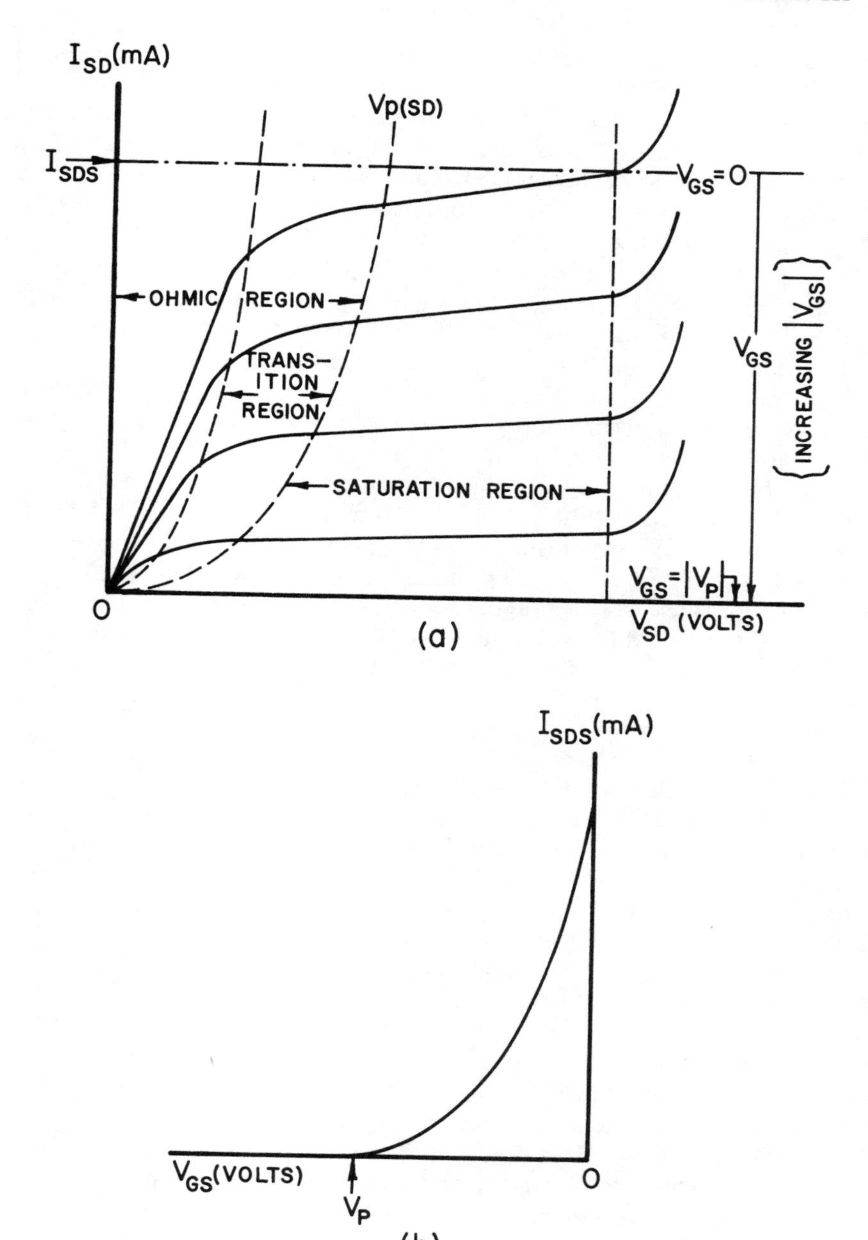

FIGURE 11-39. (a) The effect of V_{GS} upon the current-voltage characteristics of an n-type channel JFET. Note that breakdown limits the saturation region. (P-type channel JFETs have the same behavior, but the polarities are reversed.) (b) Maximum saturation current, I_{SDS}, as a function of V_{GS}. [(b) after Beeforth, T. H. and Goldsmid, H. J., *Physics of Solid State Devices,* Pion Ltd., London, 1970, 77. With permission.]

In the saturation region, JFETs may be used as sources of controlled, nearly constant currents. They may be made to act as controlled, variable resistors in the ohmic region. As such, they are very versatile devices.

Compared to BJTs, JFETs have lower "noise", high-input impedance (apparent a.c. resistance analogous to d.c. resistance.) In addition, their mechanism depends upon the flow of majority carriers and the current is affected primarily by the temperature variation of the carrier mobility. BJTs are minority carrier devices in which the number of carriers able to cross a depletion zone without recombination varies exponentially with temperature. This can lead to "thermal runaway." Thus, JFETs are less sensitive to thermal effects. The gate current does tend to increase with temperature, and steps must be taken to prevent it from decreasing the bias. The packing density of JFETs is greater in integrated circuitry than that of BJTs. Also of prime technical importance is the fact that JFETs are less complicated to manufacture.

The JFETs do not have as good a frequency response as BJTs, having useful gains up to about 0.5 GHz as compared to 5 GHz, or higher, for BJTs. The reason for this is that the width of the base may be made to be much smaller than the channel of a JFET. In addition, the gains, β in Equation 11-82, are usually much larger than G (Equation 11-100) for a given frequency range. However, the JFETs have greater linearity and stability when used in low-impedance amplifiers.

The JFET has a smaller power capacity than the BJT. This also results from the comparative size of the small base in a BJT compared to the relatively large channel in a JFET. The larger voltage drop across the channel limits the current to a greater degree than the base of a BJT, thus limiting the power capacity of the JFET.

JFETs are widely used in such applications as "square-law" amplifiers (Equation 11-98). They also are used for square-law detection and analog multipliers. Their linearity and stability are responsible for their extensive use in RF amplifiers in radio receivers. Another application is for the amplification of signals from detectors of optical radiation.

11.10.3. Metal-Oxide-Silicon FET (MOSFET)

The channel in the JFET is within the volume of one of the semi conductor members of a p-n junction. However, it is possible to induce a channel adjacent to the surface of a semiconductor.

It was shown in Section 11.2.1.1 that the unsaturated bonds of the ions adjacent to the surface of a semiconductor act as traps for electrons. The net result is a thin layer of negative charge which forms at and just below the surface. This has the effect of repelling electrons in the conduction band; it results in a volume of positively charged donor ions. And, in n-type material, the depleted zone is more like p-type than n-type material (see Figure 11-13).

When the dopant concentration is low, the number of holes which is attracted by the negative, surface charge may be larger than the number of electrons in the thin, subsurface volume. When this is the case, the surface material, for all practical purposes, may be considered to be p-type. The close proximity of the p-type surface layer to the n-type bulk material induces a depletion zone between them and the bulk material. In a similar fashion, p-type bulk material will have an n-type surface layer. Surface behavior of this kind is called surface inversion. Inversions such as these cannot occur in materials with high dopant concentrations without external assistance.

The density of surface states is very important. The early FETs were very ineffective, despite the theoretical prediction of efficient operation. Schockley and Pearson (1948) working on the basis of Bardeen's prediction (1947) showed that surface states were present and were responsible for break-down or leakage at lower voltages than for p-n junctions. In so doing, they discovered the BJT. The development of FETs was delayed until about 1953 when "cleaner" processing techniques were developed. It was not until integrated circuitry was made possible that FETs became competitive with BJTs.

The surface charge of a semiconductor also may be changed when a charge is placed

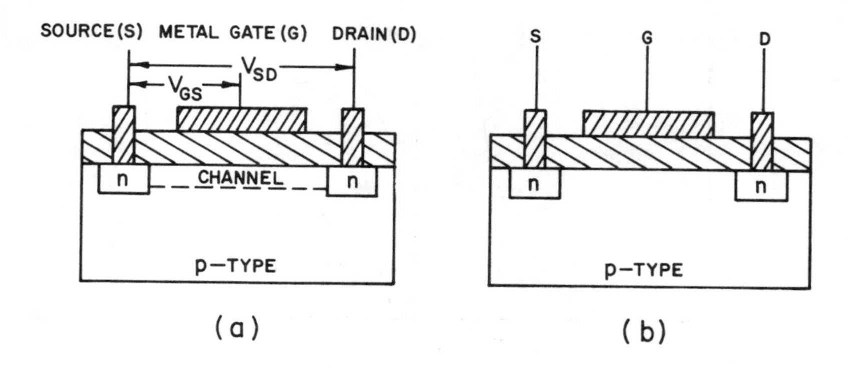

FIGURE 11-40. Schematic diagrams of MOSFETs: (a) with lightly doped p-type substrate, and showing the induced n-type channel; (b) with a heavily doped substrate. The terminals are metallic in both cases. (From Driscoll, F., Coughlin, F., *Solid State Devices and Applications,* Prentice-Hall, Englewood Cliffs, 1975, 134. With permission.)

upon a metallic electrode which is insulated from the semiconductor. A positive charge upon the electrode opposes the negative surface in the semi-conductor and diminishes the inversion layer. Conversely a negative charge upon the electrode increases the inversion layer. Inversion layers are induced in semiconductors with high dopant concentrations by this external means. Devices employing both of these techniques are called MOSFETs. They also are known as insulated-gate FETs (IGFET) or metal-insulated semiconductor transistors (MIST) (Figure 11-40).

The operating mechanism of a MOSFET is based upon the manipulation of the inversion layer between the two n-type volumes shown in the figure. In a lightly doped substrate (Figure 11-37a) conduction is possible with $V_{GS} = 0$. Variations in V_{GS} change the extent of the inversion layer; negative values of V_{GS} decrease the p-type inversion layer and the n-type depletion zone of the p-type material until pinch-off occurs. In the same way, positive values increase the width of the n-type inversion zone and I_{SD} increases. The same mechanism takes place in highly doped substrates. However, the onset of conduction requires higher values of V_{GS} as the dopant concentration increases. This is shown in Figure 11-41a. Transistors of this kind are called enhancement-type MOSFETs. Heavily doped MOSFETs cannot conduct until an inversion channel is created between the source and the drain because of the p-n junctions.

MOSFETs also are constructed with n- or p-type channels between the terminals. These are called depletion-type devices. They react in the same way as JFETs. Their current-voltage characteristics are shown in Figure 11-42. These are made in order to avoid reliance upon the highly variable surface charges.

The similarities between Figures 11-39, 11-41, and 11-42 are apparent. All of the curves show a V_P. In each case, the curves of I_{SD} vs. V_{SD} show ohmic and saturation regions. These devices also follow a square-law amplification, different from Equation 11-98, which varies linearly with V_{GS} at saturation.

MOSFETs may be used in virtually all of the applications given for JFETs, except that they have useful gains up to about 30 GHz.

Some of the materials which have been used for these devices include CdS, CdSe, GaAs, Ge, InAs, InSb, PbS, PbTe, Si, and SnO_2. One of the most common types consists of an Si substrate and an SiO_2 insulating layer.

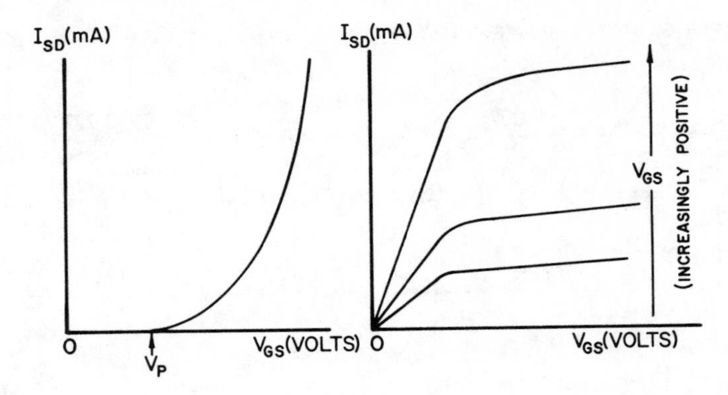

FIGURE 11-41. Current-voltage characteristics of enhancement-type MOSFETs. (After Beeforth, T. H. and Goldsmid, H. J., *Physics of Solid State Devices,* Pion Ltd., London, 1970, 79. With permission.)

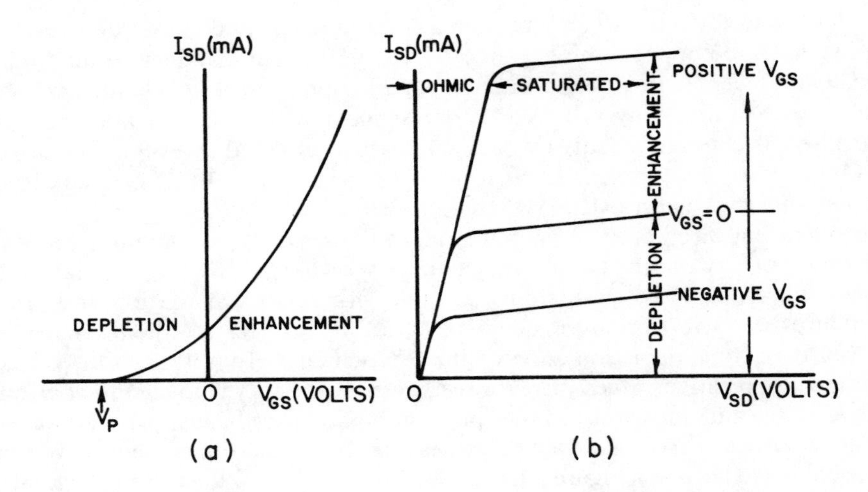

FIGURE 11-42. Effects of V_{GS} upon the current-voltage characteristics of MOSFETs, including both depletion and enhancement modes. (After Beeforth, T. H. and Goldsmid, H. J., *Physics of Solid State Devices,* Pion Ltd., London, 1970, 79. With permission.)

11.11. PROBLEMS

1. Derive Equation 11-20 from Equation 2-13.
2. Explain the small temperature dependencies of R_H for metals as compared to semiconductors.
3. How are large temperature variations expected to affect the value of R_H for semiconductors? Describe some sources for this implicit in the test method.
4. Why must one of the Hall probes be adjusted so that $V_H = 0$ for the determination of θ_H?.
5. Describe the use of R_H in the quality control of transistor manufacture.
6. Use the data in Table 11-3 to calculate the valence of copper (refer to handbooks for lattice parameter).

7. Explain the application of the Heisenberg Uncertainty Principle to the values of m^* at the discontinuities shown in Figure 11-5b.
8. Discuss and explain the band configurations shown in Figure 11-6.
9. Discuss the convenience of the use of the concept of effective mass.
10. Why do donor or acceptor ions create states within the gap?
11. Why should such a condition as the exhaustion range exist in extrinsic semiconductors?
12. Give the justification for the use of the Boltzmann tail in the determination of the number of electrons in the conduction band.
13. Explain the differences in the behaviors of the Fermi levels of semiconductors with high and low dopant concentrations as functions of temperature in Figure 11-11.
14. Explain why the dielectric constant of the host material may be used in the calculation of the ionization energy of a donor or acceptor ion, as in Equations 11-50 and 11-50a.
15. Why should a crystal imperfection constitute a trap?
16. Explain the significance of the degree of ionization of a trap upon its capacity to retain a carrier (use both classical and quantum concepts).
17. Is the assumption that N_T, in Equation 11-53, is independent of time correct? If not, what are the implications?
18. Why do the impurities C, N, and O have very little effect upon the properties of Ge and Si?
19. Why are the carrier lifetimes in commercial materials so much shorter than their theoretical values?
20. Show that the wavelength associated with leV is 1.24×10^{-4} cm.
21. Sketch and describe the change in the barrier of metal 2 (Figure 11-17) when a reverse bias is applied.
22. Sketch and describe the barrier and space-charge region at a metal-p-type semiconductor junction.
23. Explain the presence of a threshold voltage such as is shown in Figure 11-21.
24. Explain why the electric field at the surface of a semiconductor can be larger than within the bulk material.
25. Use data from Table 11-2 to calculate the temperatures of the onset of intrinsic behavior in Ge and Si.
26. Show, by means of a specific, compound semiconductor, how it may be either n- or p-type, depending upon stoichiometry.
27. Why are large, dark currents undesirable in photoconducting devices?
28. Select a material from Table 11-2 which would give a good photoelectric response to radiation with $\lambda = 0.88\ \mu m$.
29. Why may the Boltzmann equation be used to describe population inversion?
30. Describe an inefficient set of conditions for the generation of coherent radiation and explain its deficiencies.
31. Is there any advantage to doping semiconductors to be used for strain gages so that they operate in the exhaustion range?
32. Is there any advantage to doping semiconductors to be used for thermistors so that they operate in the exhaustion range?
33. Explain how the combination of FeO and Fe_2O_3 can have properties desirable for thermistors.
34. How may a doped semiconductor be employed to compensate for temperature-induced changes in electronic circuits?
35. How would you design a Peltier cooling system for a large electronic computer?
36. What is the reason for the large differences in the Hall coefficients of metals and semiconductors? Which is easier to measure? Why?

37. Sketch the sensing element, and its various leads, of a device based upon the Hall effect for use as a gaussmeter.
38. How may burn-out occur in an avalanche diode?
39. Could a series of avalanche diodes be used to calibrate a voltmeter? Explain.
40. Explain why the large, mean-free path of an electron injected into a GaAs diode is responsible for the high current densities required, as well as their low lasing amplifications.
41. Compare the capacitance capabilities of Varactors and p-i-n devices and explain their respective advantages and disadvantages.
42. Explain thermal runaway in a BJT.
43. Why should a JFET be less sensitive to temperature than a BJT?
44. Explain and compare the advantages and disadvantages of BJTs and JFETs.
45. Explain why surface inversion is not possible in heavily doped material without external influences.

11.12. REFERENCES

1. **Dekker, A. J.,** *Solid State Physics,* Prentice-Hall, Englewood Cliffs, N.J., 1957.
2. **Sproull, R. L.,** *Modern Physics,* 2nd ed., John Wiley & Sons, New York, 1963.
3. **Kittel, C.,** *Introduction to Solid State Physics,* 3rd ed., John Wiley & Sons, New York, 1966.
4. **Stringer, J.,** *An Introduction to the Electron Theory of Solids,* Pergamon Press, Elmsford, N.Y., 1967.
5. **Seitz, F.,** *Modern Theory of Solids,* McGraw-Hill, New York, 1940.
6. **Cohen, M. M.,** *Introduction to the Quantum Theory of Semiconductors,* Gordon and Breach, New York, 1972.
7. **Wolf, H. F.,** *Semiconductors,* Interscience, New York, 1971.
8. **Wilkes, P.,** *Solid State Theory in Metallurgy,* Cambridge University Press, New York, 1973.
9. **Feldman, J. M.,** *The Physics and Circuit Properties of Transistors,* John Wiley & Sons, New York, 1972.
10. **Leck, J. H.,** *Theory of Semiconductor Junction Devices,* Pergamon, Press, Elmsford, N.Y., 1967.
11. **Bylander, E. G.,** *Materials for Semiconductor Functions,* Hayden, Rochelle Park, N.J., 1971.
12. **Driscoll, F. F. and Coughlin, R. F.,** *Solid State Devices and Applications,* Prentice-Hall, Englewood Cliffs, N.J., 1975.
13. **Beeforth, T. H. and Goldsmid, H. J.,** *Physics of Solid State Devices,* Pion Ltd., London, 1970.

Chapter 12

DIELECTRIC PROPERTIES

Dielectric materials are a special class of substances which, under almost all conditions, are insulators. They have the interesting and useful property that their electrons, ions, or molecules may be polarized under the influence of an external electric field. When such materials are placed between the plates of capacitors (condensers) they increase the total capacity of these devices. This application constitutes one of the major applications of these materials.

A much smaller subset of this class of materials consists of crystals which are polarized spontaneously. Their behavior parallels that of ferromagnetic materials. These are known as ferroelectric crystals. Such materials have very high values for their dielectric constants and permit the construction of small capacitors with high capacity and other electronic devices. Ferroelectrics constitute a category of materials within the class of pyroelectrics; these materials change their polarizations with temperature.

All of the ferroelectric materials and some of the other dielectric materials are piezoelectric, that is, they can change their polarizations and dimensions under the influence of external mechanical forces or electric fields. These materials can serve as electro-mechanical transducers because a mechanical stress can induce an electric field and, conversely, an external field can induce a mechanical stress. Ferroelectric crystals and quartz crystals (which are not ferroelectric) are used both as frequency generators and as detectors of electromagnetic radiation and ultrasonic waves.

The largest engineering application of dielectric materials is for purposes of electrical insulation. A large range of widely differing materials is used for this very important engineering application.

The basic thermal properties of crystalline insulators (heat capacity, thermal expansion, and thermal conductivity) are given in Chapter 4 (Volume I). The general band structure of this class of solids is presented in Chapter 5 (Volume I). It also is noted in the previous chapter that some insulators may react like semiconductors under certain conditions.

Many organic materials which demonstrate high degrees of polarizability and excellent electrical insulating properties are noncrystalline. These materials are also considered here.

Some of the basic engineering properties of this class of materials such as dielectric constants, polarizabilities, the effects of both static and alternating fields, conduction, breakdown, and optical properties will be examined.

12.1. CAPACITORS — STATIC FIELDS

Two parallel, conduting plates close to each other, and connected in an electrical circuit, constitute a capacitor. The measure of the ability of the two plates to store an electric charge, when a voltage exists between them, is known as the capacitance of the assembly. The capacitance, C, is given by

$$C = \frac{Q}{V} \tag{12-1}$$

in which Q is the charge in coulombs and V is the voltage. The unit of capacitance, coulombs per volt, is the farad. Capacitances usually are given in units of small submultiples of the fared.

The properties of a capacitor are determined largely by the geometry of its plates. Two parallel plates each with an area A, separated by a distance d and charged with equal and opposite charges Q on each plate, will store these charges on the surfaces of the plates when d is small with respect to A.

The homogeneous electric field strength between the plates when a vacuum is present between them is

$$\bar{E}_v = 4\pi q = D \qquad (12\text{-}2)$$

where q is the surface charge density. D is known as the flux density, the electric induction, or the displacement. D is a measure which depends upon the strength and distribution of the charges which produce the field. The potential difference between the plates, when a vacuum exists between them, is

$$V_v = \bar{E}_v d$$

The capacitance of the assembly is given by

$$C_v = \frac{Aq}{V_v} = \frac{Q}{V_v} \qquad (12\text{-}3)$$

If a dielectric material is placed between the plates, its electrical effect is to increase the capacitance of the assembly. This is a result of the polarization of the molecules of the intervening material. Substances composed of polar molecules (those with a permanent electric dipole moment) have random molecular orientations in the absence of an external electric field. The field between the charged plates aligns, or polarizes, the molecular dipoles. This tendency of molecular alignment to become parallel to the field is called orientation polarization. If the intervening, insulating substance is a liquid or a gas, as is sometimes the case, the polar molecules can move much more freely and become aligned more readily than in most solids. The molecules of the insulating material become predominantly arrayed such that their negative poles at one of its surfaces are adjacent to the positive poles on the surface of the positive plate. The dipolar molecules in the interior of the dielectric material array themselves in a −, +, −, + order until, at the opposite surface of the material, the positive molecular poles are adjacent to the negative plate. A layer of negative charge thus forms on the surface of the dielectric adjacent to the surface of the positive plate and one of positive charge is created adjacent to the surface of the negative plate. The result is the effective cancellation of some of the charges upon each of the plates and the consequent reduction of the electric field strength. This increases the ability of the plates to hold more charges; their capacity is increased. In cases where the molecules are not polar (those with symmetric charge distributions) polarization may be induced by the applied field to produce results similar to those just noted. This can occur in solids where the molecules or ions are not free to rotate.

It should be noted that orientation polarization shows a temperature dependence similar to that described in Section 8.3.1. (Volume II) (see Section 12.3). The polarization decreases with increasing temperature. Nonpolar materials (Section 12.2) are less affected by temperature. In this case, the change in properties is largely a result of the change in the intermolecular distance, changes in bond angles and/or the shift of the electron charge distributions around the nuclei of the molecular component.

The change in the properties of a capacitor caused by the presence of the dielectric between the plates permits the definition of the static, or relative, dielectric constant ε. This also is known as the electric permittivity. It is a dimensionless number. This is based upon Equation 12-3 and uses the case of the vacuum between the plates as a reference. Thus,

$$\epsilon = \frac{V_V}{V} = \frac{C}{C_V} \tag{12-4}$$

The reduction in the field strength caused by the presence of the dielectric material between the plates may be expressed as

$$\bar{E}_V = \epsilon\bar{E} = D; \ \epsilon = \frac{D}{\bar{E}} \tag{12-5}$$

Here $\bar{E}$ is the applied field when the dielectric is present between the plates.

The reduced charge density upon the surface of the plates resulting from the presence of a dielectric material between them is

$$q' = \frac{\bar{E}}{4\pi} \tag{12-6}$$

The reduction of the surface charge on the plates due to the presence of the intervening material is the polarization P. The resulting increased capacitance is given by $C = (Q + P)/V$. Thus, the polarization is the difference between the two conditions:

$$P = q - q' = \frac{\bar{E}_V}{4\pi} - \frac{\bar{E}_D}{4\pi} \tag{12-7}$$

where $-\bar{E}_D/4\pi$ represents the depolarization of the plates by the dielectric. This can be reexpressed using Equation 12-5 as

$$P = \frac{\bar{E}_V}{4\pi} - \frac{\bar{E}_V}{4\pi\epsilon} = \frac{\bar{E}_V}{4\pi}\left[1 - \frac{1}{\epsilon}\right]$$

or as

$$P = \frac{\bar{E}_V}{4\pi\epsilon}(\epsilon - 1) \tag{12-8}$$

Rearrangement and further use of Equation 12-5 also gives

$$P = \frac{\bar{E}}{4\pi}(\epsilon - 1) \tag{12-9}$$

The polarization, P, obtained in this way is the polarization per unit volume; it is discussed in detail in following sections. Rearranging Equation 12-9 gives

$$\epsilon = \frac{4\pi P}{\bar{E}} + 1 \tag{12-10}$$

Multiplying both sides of Equation 12-10 by $\bar{E}$ and using Equation 12-5 results in

$$\bar{E}\epsilon = D = 4\pi P + \bar{E} \tag{12-11}$$

The relationships given here include those basic properties of dielectric materials of greatest use here. Frequent recourse will be made to them in subsequent discussion.

12.2. POLARIZATION OF ATOMS AND MOLECULES

In considering molecular polarization it is helpful to start with an examination of the behavior of isolated ions or molecules so that any neighbor-neighbor reactions can be eliminated. An example of this would be a gas at low pressure.

In neutral systems, two charges of $+e$ and $-e$ a distance x apart have a dipole moment of $p = ex$. The dipole moment of a set of such dipoles is

$$P = \sum_j ex_j = Np \tag{12-12}$$

in which N is the number of dipoles per unit volume.

The simplest case is that of an isolated atom, or ion. In the absence of an external field, the centroid of the electric charge distribution is at the nucleus. The positive and negative charges may be considered to act as though they are superimposed. The atom or ion has no dipole moment in this situation. The application of a homogeneous, static field can cause the center of the electron charge to become offset from its initial position. This elastic displacement will be in a direction opposite to the applied field in a manner analogous to that shown in Figure 5-10 b (Volume I). The attractive forces between the nucleus and the electrons counteract the displacement of the electrons caused by the field. An equilibrium is reached between these two effects and the resulting offset between the centers of the two charges results in a dipole moment. This induced dipole moment is given approximately by

$$P_a = \alpha_e \bar{E} \tag{12-13}$$

where α_e is the electronic polarizability of the atom or ion.

A simple model may be used to estimate the magnitude of α_e. Picture the particle under consideration as being composed of a uniform, negative, spherical electron cloud of radius r surrounding a nucleus of charge Ze, where Z is the number of electrons. The application of an electric field will offset the centers of charge by an amount x. The attractive forces between the nucleus and the electrons counteract the displacement resulting from the field. The force exerted by the field is

$$F = Ze\bar{E}$$

Using Gauss's law, the field at a distance x from the center of a spherical field is

$$\bar{E} = \frac{(Ze)'}{x^2}$$

For points within the spherically uniform charge distribution of radius r

$$(Ze)' = Ze \frac{4/3\,\pi x^3}{4/3\,\pi r^3} = Ze(x/r)^3$$

At equilibrium the restoring force must equal the force exerted by the field so

$$F = Ze\bar{E} = Ze\frac{(Ze)'}{x^2} = Ze\frac{Ze(x/r)^3}{x^2} = \frac{(Ze)^2 x}{r^3} \tag{12-14}$$

and

Table 12-1
SCREENING CONSTANTS AND
COEFFICIENTS FOR EQUATION 12-15a

Shielded electron	Shielding electron 1s	2s	2p	3s	3p	n	A
1s	0.35	0.00	0.00	0.00	0.00	1	4.5
2s	0.85	0.35	0.35	0.00	0.00	2	1.1
2p	0.85	0.35	0.35	0.00	0.00	3, etc.	0.65
3s	1.00	0.85	0.85	0.35	0.35		
3p	1.00	0.85	0.85	0.35	0.35		

From Slater, J. C. and Frank, N. H., *Introduction to Theoretical Physics,* McGraw-Hill, New York, 1933, 432. With permission.

$$Ze\bar{E}r^3 = (Ze)^2x; \quad \bar{E}r^3 = Zex \tag{12-14a}$$

From Equations 12-13 and 12-14a the electronic polarization is

$$p_e = \alpha_e\bar{E} = Zex = r^3\bar{E} \tag{12-15}$$

From this it is seen that $\alpha_e = r^3$. Since r (see Table 10-3) is in angstroms, α_e is approximately 10^{-24} cm^3. The amount of offset of the charge depends upon the strength of the field. When an external field is applied, x will be very small; the charge upon an electron being 1.6×10^{-19} coul (4.8×10^{-10} esu), x will be about 10^{-15} E. Thus, x will be very much less than r. In alternating fields the electronic polarizability is essentially constant up to ultraviolet frequencies (see Section 12.6 and Figure 12-7).

The electronic polarizability may be calculated (based upon Equation 12-15 by an empirical relationship given by Slater and Frank.[15] This is, for each electron in the outer level,

$$\alpha_e = r^3 \simeq A\left[\frac{n^2 r_o}{Z - S}\right]^3 \tag{12-15a}$$

where n is the principal quantum number of the electrons in the highest filled level, r_o is the Bohr hydrogen radius (0.53 Å), and Z is the atomic number. The coefficient A and the screening constants, S, are given in Table 12-1.

Another empirical relationship may be used to approximate α_e. This is given by

$$\alpha_e \simeq \frac{(r \pm 0.85)^3}{8 - n} \tag{12-15b}$$

Here r is rCN12 for cations as given in Table 10.3. In this case, the numerator is $(rCN12-0.85)^3$. For the case of anions the numerator is $(rCN6 + 0.85)^3$. In both cases, n is the same as that for Equation 12-15a. The average error of Equation 12-15b is approximately ±40%. However, when elements with atomic numbers greater than 10 are considered, the average error is approximately ±20%. These errors are equal to or less than those given by Equation (12-15a).

Every ion will have a given polarizability. The larger the number of electrons that an atom has, the larger its electronic polarizability will be. Electrons in tightly bound

Table 12-2
POLARIZABILITIES

A. Some Electronic Polarizabilities
(10^{-24} cm^3, Sodium D Lines)

		H	0.67				
		He	0.20	Li^+	0.03	Be^{+2}	0.008
F^-	1.05	Ne	0.40	Na^+	0.18	Hg^{+2}	0.09
Cl^-	3.69	Ar	1.64	K^+	0.84	Ca^{+2}	0.47
Br^-	4.81	Kr	2.48	Rb^+	1.42	Sr^{+2}	0.86
I^-	7.16	Xe	4.16	Cs^+	2.44	Ba^{+2}	1.56
H_2	0.80	HCN	2.58	F_2	1.16	P_4	14.71
HCl	2.64	H_2O	1.49	Cl_2	4.60	BCl_3	8.31
HBr	3.62	NH_3	2.23	Br_2	6.90	BBr_3	11.87
HI	545	H_2S	3.80	N_2	1.76	CCl_4	10.53

B. Some Polarizabilities of Bond Types in Organic Molecules

H−	0.40	F−	0.58	
C in CH_4	1.03	Cl−	2.30	Three-membered ring 0.24
C in diamond	0.83	Br−	3.45	
>C−	1.34	I−	5.50	Four-membered ring 0.13
−C≡	1.42	O<	0.69	
N< (three bonds)	0.97	O=	0.85	
−N=	0.90	S<	3.00	
N≡	0.83	S=	3.14	

From Pauling, L., *General Chemistry*, 3rd ed., W. H. Freeman, San Francisco, 1970, 397. With permission.

inner shells have smaller effects than those in outer levels, which are less tightly bound. Metal atoms in the ground state have higher polarizabilities than when ionized; the valence electrons are absent in ionized states and leave the nuclei surrounded by relatively tightly bound, filled shells. Negative ions show opposite behavior. Some electronic polarizabilities are given in Table 12-2.

Molecules show more complex behaviors than atoms or ions. Some of these may have permanent dipoles, depending upon their charge distributions. These dipoles are affected by the application of external electric fields.

Where the electronic charge distribution within a molecule is symmetric, it has no permanent dipole. However, the application of an external electric field may change the interionic spacings and/or the bond angles and induce a dipole moment. Such elastic shifts in molecular configurations give rise to dipole moments which are described by

$$p_a = \alpha_a \bar{E}_{loc} \tag{12-16}$$

where α_a is the atomic polarizability and E_{loc} is the local, or internal, field surrounding it (see Section 12.4). The atomic polarizability is the average effect of all possible polarizations in relation to the applied field. This property is essentially constant up to infrared frequencies (see Figure 12.7). Equation 12-16 leads to an expression for the atomic polarization per unit volume given by

$$P_a = Np_a = N\alpha_a\bar{E}_{loc} \quad (12\text{-}17)$$

where N is the number of dipoles per unit volume. These effects are discussed in subsequent sections.

Permanent dipoles are created in molecules when the electronic charge distributions are nonsymmetrical. This may be illustrated by considering a simple case. Where two atoms with relatively large differences in electronegativity form a molecule, the valence electrons are more closely associated with the atom which has the higher electronegativity; a more negative charge is associated with it. Consequently, a more positive charge is associated with the other atom. The result is the creation of a dipole and the molecule has a dipole moment (Equation 12-12). The distance between the centers of positive and negative charges provide a measure of the degree of asymmetry of molecules.

Dipole moments are measured in Debye units. This unit is derived by considering a dipole in which each of the charges are separated a distance $x = 10^{-8}$ cm (1 Å) and are equal to $|4.8 \times 10^{-10}|$ esu, the charge on an electron. Thus, the Debye unit is $D = 10^{-8} \times 4.8 \times 10^{-10} = 4.8^{-18}$ esu-cm.

When the dipole moment and the distance between the charges on a molecule are known, the charges on the dipole, e', can be determined from Equation 12-12 as $e' = p_d/x$. This approximates the amount of charge participating in the dipole. Thus, e' and x can provide insight into molecular structure.

On the basis of the foregoing, more specific explanations may be made regarding the relation of polarity to molecular structure. Starting with the simplest case, the noble gases are monatomic, have symmetrical structures, and have no permanent dipole moment. The molecular gases of elements are diatomic, symmetrical, and are, thus, nonpolar. Diatomic molecules of composition AB are asymmetric, as noted above, and therefore polar. Molecules of composition AB_2 are non-polar if their molecular structures are linear and A is centrally located between both B atoms. However, if A is asymmetrically positioned, the molecule will be polar. Also, when the two B atoms form an angular array with the A atom, the molecule will be polar and $p_d = 2p \cos \theta/2$, where p is the single-bond moment and θ is the angle between the bonds formed by the two B atoms and the A atom. When the bond angles of AB_3 molecules are equal and coplanar, no dipole moment can be present. Such molecules will be polar when these conditions are not met. One such case is where AB_3 molecules form tetrahedra. The lack of symmetry caused by the A atom will cause such molecules to be polar. Where AB_4 molecules form either equal-angle, coplanar bonds, or where they form tetrahedra with the A atom at the centroid, as in Figure 10-17a, no dipole moment will be present. Examples of molecules of the types described here are given in Table 12-3.

As previously noted, molecules with permanent dipoles will tend to become oriented parallel to an applied electric field. This behavior is known as orientation polarization and is described in the next section.

12.3. ORIENTATION POLARIZABILITY

Substances whose molecules have permanent dipole moments and which are relatively free to rotate are considered here. These materials include many gases and liquids. The following theory is based essentially upon the Langevin model given in detail in Section 8.3.1 (Volume II). Debye used the results of this approach to include the contributions of permanent dipoles to the total polarization (see Section 12.4).

Table 12-3
DIPOLE MOMENTS OF SOME COMPOUNDS
(In Debyes)

HCl	1.03	CH_3OH	1.68
HBr	0.78	C_2H_5OH	1.70
HI	0.38	CH_3NH_2	1.30
H_2O_2	2.10	$C_2H_5NH_2$	1.30
NH_3	1.46	CH_3Cl	2.00
H_2O	1.84	CH_2Cl	1.60
H_2S	1.10	$CHCl_3$	1.10
SO_2	1.60	C_6H_5Cl	1.73
HCN	2.93	$C_6H_5NO_2$	4.23
CH_3CN	3.20	C_6H_5OH	1.70
C_2H_5CN	3.40	$C_6H_5NH_2$	1.56

Note: Symmetric molecules such as CCL_4, C_6H_6, CO_2, CS_2, H_2, and N_2 have no dipole moments.

From Sinnott, M. J., *The Solid State for Engineers*, John Wiley & Sons, New York, 1958, 189. With permission.

In the absence of an external field, substances of the kind under discussion here will consist of randomly oriented, permanent, molecular dipoles with zero net dipole moment. The application of an external field will exert a torque on the molecular dipoles (see Figure 8-4, Volume II.) The molecules will tend to become aligned parallel to the field. Some of the alignment is destroyed by the thermal activity of the molecules as described in Section 8.3.1. The Langevin theory is used to determine the average molecular dipole moment parallel to the applied field as a function of temperature.

The potential energy of an ion or a molecule with a dipole moment p_d in an electric field, $\overline{E}$, where ϕ is the angle between p_d and E is

$$P.E. = p_d\overline{E}\cos\phi \qquad (12\text{-}18)$$

The average dipole moment per unit volume is

$$P_d = Np_d < \cos\phi > \qquad (12\text{-}19)$$

Where N is the number of ions or molecules per unit volume. The factor $< \cos\phi >$ was found to be given by

$$< \cos\phi > = L(a) = \coth a - \frac{1}{a} \qquad (8\text{-}46)$$

It was noted previously that when a is greater than 5, changes in L(a) become relatively small. This corresponds to the greatest degree of alignment of the dipoles. A solution was obtained for the case where $a \ll 1$. Here $L(a) = a/3$ (Equation 8-48, Volume II). In the present case $a = p_d\overline{E}/k_BT \ll 1$, so that

$$L(a) = \frac{1}{3}\frac{p_d\overline{E}}{k_BT}$$

Thus,

$$P_d = Np_d < \cos\phi > = Np_d \cdot \frac{1}{3}\frac{p_d\overline{E}}{k_BT} = \frac{Np_d^2\overline{E}}{3k_BT} \quad (12\text{-}20)$$

From this, it is seen that the molecular, or dipolar, orientation polarizability is given by

$$\alpha_d = \frac{p_d^2}{3k_BT} \quad (12\text{-}21)$$

In addition to the orientation polarizability, the total polarizability of a substance must include the electronic contribution (Equation 12-13) and the atomic polarizability (Equation 12-17). Thus,

$$\alpha = \alpha_e + \alpha_a + \alpha_d \quad (12\text{-}22)$$

The temperature dependence of the dielectric constants of such liquids as water and alcohol was explained by assuming that these molecules have different polarizabilities and, therefore, different dipole moments. The ideas of Debye, Onsager, and others along these lines provided a means for greater comprehension of molecular structure as well as dielectric behavior (see Sections 12.4 and 12.5).

12.4. THE LORENTZ INTERNAL FIELD AND THE DEBYE EQUATION

A particular atom or molecule in a liquid or a solid is affected by both the applied field and the fields of the dipoles present on other atoms or molecules surrounding it. The influence of the surrounding dipoles must be considered because of the long-range Coulomb forces. The task here is to describe the field at the location of the particle and to determine its effect upon the particle. This field is the local, or internal, field.

The local field is determined at the center of a small, spherical zone within the dielectric. This spherical volume is chosen to be sufficiently small with respect to the dimensions of the dielectric material so that the external volume surrounding the sphere may be regarded as being homogeneous and continuous and have a dielectric constant ε. However, the sphere must be large with respect to the dimensions of a molecule. This permits the discrete nature of the material within the spherical volume to be taken into consideration as shown in Figure 12-1.

The internal field is affected by several factors. The first of these is the influence of the displacement, D, or the charge density on the plates on either side of the dielectric, as given by Equation 12-2. In addition, the effect of the charges induced at the interface between the plates and the dielectric adds the depolarizing field $\overline{E}_D = -4\pi P$ (Equation 12-7) to the field strength. The induced charges on the surface of the spherical region constitute an additional factor which also must be considered. Here it is necessary only to account for the components parallel to the applied field. An element of surface area, which is an infinitessimal ring, on the surface of the spherical region has an area of $2\pi r^2 \sin\theta d\theta$. The angle θ determines the surface charge density which is given by $-P\cos\theta$. Then, $-P\cos\theta \cdot 2r^2 \sin\theta d\theta$ is the charge on the element of surface area. This gives an electrostatic field at the center of the spherical region, parallel to the applied field. The field on the spherical surface is

$$d\overline{E}_s = \frac{\text{area} \times \text{charge density} \times \cos\theta}{(\text{distance})^2}$$

or

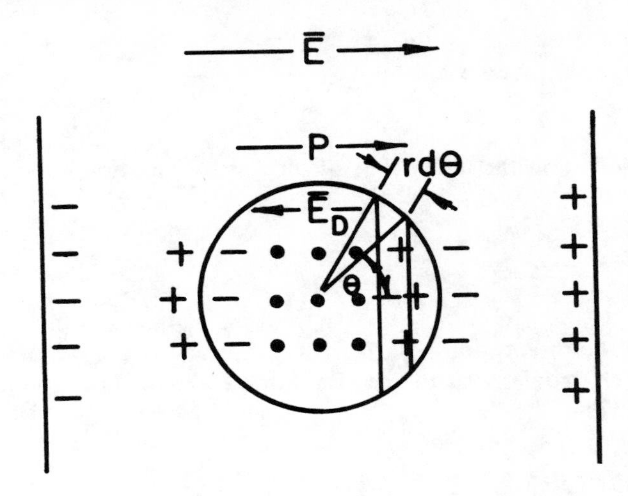

FIGURE 12-1. Basis for examining the internal field within a spherical region contained in a dielectric material which is assumed to be homogeneous and continuous.

$$d\bar{E}_s = -\frac{2\pi r^2 \sin\theta d\theta \cdot P\cos\theta \cdot \cos\theta}{r^2}$$

The charge on the spherical surface thus is found, in the same manner as for Equation 4-146 (Volume I), to be

$$\bar{E}_s = -2\pi P \int_0^\pi \cos^2\theta \sin\theta d\theta = \frac{4\pi}{3} P \qquad (12\text{-}23)$$

The sum of these three components ($D + E_D + E_s$) gives the local field as

$$\bar{E}_{loc} = D - 4\pi P + \frac{4\pi}{3} P = D - \frac{8\pi}{3} P \qquad (12\text{-}24)$$

Using Equation 12-11 for D

$$\bar{E}_{loc} = 4\pi P + \bar{E} - \frac{8\pi}{3} P = \bar{E} + \frac{4\pi}{3} P \qquad (12\text{-}25)$$

The substitution of Equation 12-9 into Equation 12-25 gives

$$\bar{E}_{loc} = \bar{E} + \frac{4\pi}{3} \cdot \frac{\bar{E}}{4\pi} (\epsilon - 1) = \bar{E}[1 + 1/3(\epsilon - 1)]$$

$$\bar{E}_{loc} = \bar{E}\,\frac{\epsilon + 2}{3} \qquad (12\text{-}26)$$

Equation 12-26 is the Lorentz field. Equations 12-25 and 12-26 do not include the effects of the dipoles within the spherical region. The field within the spherical region depends upon the crystal lattice type, since the discrete nature of the material must be considered. When the lattice type is simple cubic, FCC, or BCC, the net effects of parallel dipoles on the lattice within the spherical region are zero. This is a result of the symmetry of the dipoles about the center of the sphere; a dipole with a given coordinate is balanced by another dipole with the mirror coordinate of the first one. In complex cubic lattices, this is not always the case. Barium titanate has a cubic lattice, but the oxygen atoms do not have cubic symmetry; these influence the local field and a net nonzero moment results (see Section 12.11).

The results given by Equation 12-26 may not always agree with experiment even when the crystal structure is one of the cubic types. This can result from nonhomogeneous fields induced by overlapping levels or by near neighbors. Where a crystal is composed of several different atoms, each atom will have a different local field because each will have different surroundings. The internal field at the site of one such atom is given by a modification of Equation 12-25 as

$$\bar{E}_{loc} = \bar{E} + \beta P \qquad (12\text{-}27)$$

Here β represents the particular field constant for the given case. Thus, Equation 12-25 is a special case of Equation 12-27; the former holds only when the net effects of the dipoles within the spherical region are zero.

Equation 12-26 provides a basis for the derivation of the Debye equation when used with Equation 12-9. The latter equation may be written as

$$P = \frac{\bar{E}}{4\pi}(\epsilon - 1) = P_a + P_d$$

Equations 12-17 and 12-20 are substituted for P_a and P_d to give

$$P = \frac{\bar{E}}{4\pi}(\epsilon - 1) = N\alpha_a\bar{E}_{loc} + N\frac{p_d^2}{3k_BT}\bar{E} = N\left[\alpha_a\bar{E}_{loc} + \frac{p_d^2}{3k_BT}\bar{E}\right] \qquad (12\text{-}28a)$$

Equation 12-26 is substituted for both of the field factors in the brackets giving

$$P = \frac{\bar{E}}{4\pi}(\epsilon - 1) = N\bar{E}\,\frac{\epsilon + 2}{3}\left[\alpha_a + \frac{p_d^2}{3k_BT}\right]$$

The fields vanish and, upon simplification, the relationship is found to be

$$P = \frac{\epsilon - 1}{\epsilon + 2} = \frac{4\pi}{3}N\left[\alpha_a + \frac{p_d^2}{3k_BT}\right] \qquad (12\text{-}28b)$$

This is the Debye equation for a pure substance. It combines the Langevin theory (Equation 12-20) and, as will be seen, the Clausius-Mossotti equation (Equation 12-38). In so doing, it takes the effects of permanent dipoles and temperature into consideration.

It is helpful to define a molar polarization as

$$P_m = \frac{\epsilon - 1}{\epsilon + 2}\cdot\frac{M}{\rho} \qquad (12\text{-}29)$$

in which M is the molecular weight and ϱ is the density (see Equation 12-38a). The use of Equation 12-28b gives the Debye equation as

$$P_m = \frac{4\pi}{3}N\left[\alpha_a + \frac{p_d^2}{3k_BT}\right] \qquad (12\text{-}30)$$

in which, because of the definition of Equation 12-29, N must be Avogadro's number.

The Debye equations give erroneous results for concentrated polar liquids because the Lorentz internal field does not account for short-range, molecule-molecule associations which may increase or decrease the dipole moment. However, it may be used for dilute solutions of polar molecules in nonpolar liquids, or for gases at normal pressures. A correction may be used to avoid these limitations. This is included in Equation 12-28b as

$$P = \frac{\epsilon - 1}{\epsilon + 2} = \frac{4\pi}{3} N \left[\alpha_a + \frac{p_d^2}{3k_B T + C p_d^2} \right]$$

where C is a constant which ranges between 1.0 and 1.6 for compounds which do not involve hydrogen bonds. This correction is explained on the basis that part of the field aligning the dipoles, E_d, is not taken into account in this derivation (see Equation 12-49).

12.5. THE CLAUSIUS-MOSSOTTI AND ONSAGER EQUATIONS

As noted previously, the dielectric polarization can be considered to arise from three major sources: electronic, ionic/atomic, and dipolar polarizations. Thus,

$$P = P_e + P_a + P_d \tag{12-31}$$

This relationship provides a means for differentiating among various important groups of dielectrics. One type represents the case in which the principle effect is electronic and the ionic/atomic and dipolar components are negligible or zero. Another set consists of those materials in which the contributions are from electronic and ionic/atomic sources with no dipolar component. A third class is one which includes components arising from the three sources given in Equation 12-31. One method for describing the case for $P = P_a + P_d$ is shown by the Debye approach in the previous section.

12.5.1. The Clausius-Mossotti Equation

A local field is assumed for substances in which the main polarization effect is electronic. Then, from Equations 12-17 and 12-9

$$P_e = N\alpha_e \overline{E}_{loc} = \frac{\overline{E}}{4\pi} (\epsilon - 1) \tag{12-32}$$

where N is the number of dipoles per unit volume. An expression for $\overline{E}_{loc}$ may be obtained by means of Equations 12-27 and 12-17 as applied to electronic polarization.

$$\overline{E}_{loc} = \overline{E} + \beta P = \overline{E} + \beta N \alpha_e \overline{E}_{loc}$$

Collecting terms,

$$\overline{E}_{loc} (1 - \beta N \alpha_e) = \overline{E} \tag{12-33}$$

Equation 12-32 is multiplied by 4π and Equation 12-33 is substituted into it for $\overline{E}$. The result is, since $\overline{E}_{loc}$ vanishes,

$$4\pi N \alpha_e = (1 - \beta N \alpha_e)(\epsilon - 1)$$

or

$$\epsilon - 1 = \frac{4\pi N\alpha_e}{1 - \beta N\alpha_e} \qquad (12\text{-}34)$$

In this case $\beta = 4\pi/3$ since the effects of the dipoles in the spherical Lorentz region are taken as zero (Equations 12-25 and 12-27).

Another useful relationship can be obtained by starting with Equation 12-25 rearranged as

$$\bar{E} = \bar{E}_{loc} - \frac{4\pi}{3} P \qquad (12\text{-}35)$$

Both sides of Equation 12-32 are divided by $\bar{E}$ and Equation 12-35 is substituted in the denominator to give

$$\chi_e = \frac{P_e}{\bar{E}} = \frac{N\alpha_e \bar{E}_{loc}}{\bar{E}} = \frac{N\alpha_e \bar{E}_{loc}}{\bar{E}_{loc} - \frac{4\pi}{3} P} \qquad (12\text{-}36)$$

where χ_e is the electric susceptibility. Another expression for χ is obtained from Equation 12-9. This is equated to Equation 12-36 in which Equation 12-17 is substituted for P. The result is

$$\chi_e = \frac{P}{\bar{E}} = \frac{1}{4\pi}(\epsilon - 1) = \frac{N\alpha_e \bar{E}_{loc}}{\bar{E}_{loc} - \frac{4\pi}{3} N\alpha_e \bar{E}_{loc}}$$

or, since $\bar{E}_{loc}$ vanishes,

$$\chi_e = \frac{P}{\bar{E}} = \frac{\epsilon - 1}{4\pi} = \frac{N\alpha_e}{1 - \frac{4\pi}{3} N\alpha_e} \qquad (12\text{-}37)$$

Then, reexpressing Equation 12-37,

$$(\epsilon - 1)\,[1 - (4\pi/3)\,N\alpha_e] = 4\pi N\alpha_e$$

Upon multiplication and simplification:

$$\epsilon - \frac{4\pi\epsilon}{3} N\alpha_e - 1 + \frac{4\pi}{3} N\alpha_e = 4\pi N\alpha_e$$

$$\epsilon - 1 = 4\pi N\alpha_e + \frac{4\pi\epsilon}{3} N\alpha_e - \frac{4\pi}{3} N\alpha_e$$

$$\epsilon - 1 = \frac{4\pi}{3} N\alpha_e\,(3 + \epsilon - 1) = \frac{4\pi}{3} N\alpha_e\,(\epsilon + 2)$$

And finally, more generally, where $\alpha = \alpha_e + \alpha_a$,

$$P = \frac{\epsilon - 1}{\epsilon + 2} = \frac{4\pi}{3} N\alpha \qquad (12\text{-}38)$$

Or, using the definition given by Equation 12-29,

$$P_m = \frac{4\pi}{3} N\alpha \qquad (12\text{-}38a)$$

Equation 12-38 is the Clausius-Mossotti equation for a pure substance. It relates the polarizability and the dielectric constant. It will be noted that this equation does not take orientation effects into account and is not a function of temperature, since α_e and α_a are independent of temperature; both are assumed to respond instantaneously to the application of an external field (see Sections 12.7 and 12.8). The atomic/ionic polarizability is constant up to infrared frequencies; α_e is constant up to ultraviolet frequencies. This makes possible the use of the Maxwell relationsip between the dielectric constant and the index of refraction:

$$\epsilon_\infty = n^2 \qquad (12\text{-}39)$$

in which n is the index of refraction of a substance; ε_∞ is the high-frequency component of the dielectric constant which results from the tightly bound electrons.

The substitution of n^2, or ε_∞, for ε in Equation 12-38 gives the Lorenz-Lorentz equation. One advantage of the use of Equation 12-39 in Equation 12-38 is that the refractive index may be measured more accurately than the dielectric constant in this range of frequencies. The dielectric constant is a function of both α_a and α_e up to infrared frequencies, but is only a function of α_e in the range of frequencies from infrared to ultraviolet. This is the case because α_a and α_d are essentially zero since the dipoles they represent cannot respond quickly enough to the very rapid changes in the field in this range of high frequencies. Equations 12-38 and 12-39 provide a useful experimental tool in high-frequency alternating fields. It must be noted that when ε and $n^2 = \varepsilon_\infty$ are used together (as in Equation 12-68) it is important that both be measured at the same frequency.

Nonpolar polymers, such as polystyrene, polyethylene, and polytetrafluorethylene, are widely used because of their low loss factors over very extended frequency ranges (see Sections 12.7, 12.8 and 12.13). Experiments have shown that the values of ε of these materials are linear functions of their densities. This is verified by means of Equations 12-29 and 12-38a to give

$$P_m = \frac{\epsilon - 1}{\epsilon + 2} \cdot \frac{M}{\rho} = \frac{4}{3}\pi N\alpha \qquad (12\text{-}38b)$$

Or, as a function of density

$$\frac{\epsilon - 1}{\epsilon + 2} = \rho \cdot \frac{4\pi N\alpha}{3M} = \rho C \qquad (12\text{-}38c)$$

where C is a constant for a given material since N is Avogadro's number. The constants C may be calculated from published values of P_m, or from those which have been obtained from the Lorenz-Lorentz equation.

The behavior of mixtures of nonpolar substances may be obtained by writing Equation 12-38 as

$$\frac{\epsilon - 1}{\epsilon + 2} = \frac{4\pi}{3} \sum_j N_j \alpha_j \qquad (12\text{-}40)$$

Again, using Equation 12-38

$$\alpha = \frac{3}{4\pi N} \cdot \frac{\epsilon - 1}{\epsilon + 2} \tag{12-41}$$

It should be kept in mind that $\alpha = \alpha_e + \alpha_e$ or $\alpha = \alpha_e$, depending upon the frequency at which ε or ε_∞ are determined. The dielectric constant of the mixture may be calculated from Equations 12-40 and 12-41 as

$$\frac{\epsilon_{mix} - 1}{\epsilon_{mix} + 2} = \frac{4\pi}{3} \sum_j \left[N_j \frac{3}{4\pi N} \frac{\epsilon - 1}{\epsilon + 2} \right] = \sum_j f_v \frac{\epsilon_j - 1}{\epsilon_j + 2} \tag{12-42}$$

in which f_v is the volume fraction per cm^3 of the j-th component of the mixture, and ε_j is the dielectric constant of that component. Equation 12-42 generally gives very good agreement with the experimental values of the dielectric constants of mixtures of nonpolar, organic liquids (Figure 12-2). However, in some cases Equation 12-42 appears to give values for ε_{mix} which consistantly are very slightly smaller than the experimental results. It has been considered that this might be the result of the formation of some weakly polar molecular complexes in such mixtures.

12.5.2. The Onsager Equation

Onsager assumed that the particles of which the pure materials consist are spherical. No charge fluctuations (variations in the field induced by the thermal motions of the molecules which cause changes in the induced moments) are considered to occur within a particle. In addition, no interactions between particles, such as molecular alignments resulting from short-range molecular forces, are assumed to take place.

It is further assumed that the radius, r, of the hollow spherical volume, or cavity, is given by

$$(4/3)\, \pi N r^3 = 1; \; N = 3/(4\pi r^3) \tag{12-43}$$

This approach differs from that of Lorentz (Figure 12-1) in that the spherical cavity contains just one particle. In many cases, it is helpful to approximate r as being the "molecular radius". Using the Lorenz-Lorentz form of Equations 12-38 and 12-43 for N gives

$$\frac{\epsilon_\infty - 1}{\epsilon_\infty + 2} = \frac{4\pi}{3} \frac{3}{4\pi r^3} \alpha = \frac{\alpha}{r^3} \tag{12-44}$$

The Onsager model takes more factors into account within the spherical volume than does the Lorentz approach. The field of the single dipole in the hollow, spherical volume has the effect of polarizing the surrounding material. This change in the polarization of the surrounding material induces a field at the dipole, known as the reaction field $\overline{R}$. $\overline{R}$ has the same direction and is proportional to the dipole moment:

$$\overline{R} = f_R p_d \tag{12-45}$$

in which f_R is the reaction field factor and is defined for use here as

$$f_R = \frac{1}{r^3} \frac{2(\epsilon - 1)}{2\epsilon + 1} \tag{12-46}$$

An additional dipole $\alpha\overline{R}$ is induced by the reaction field. So,

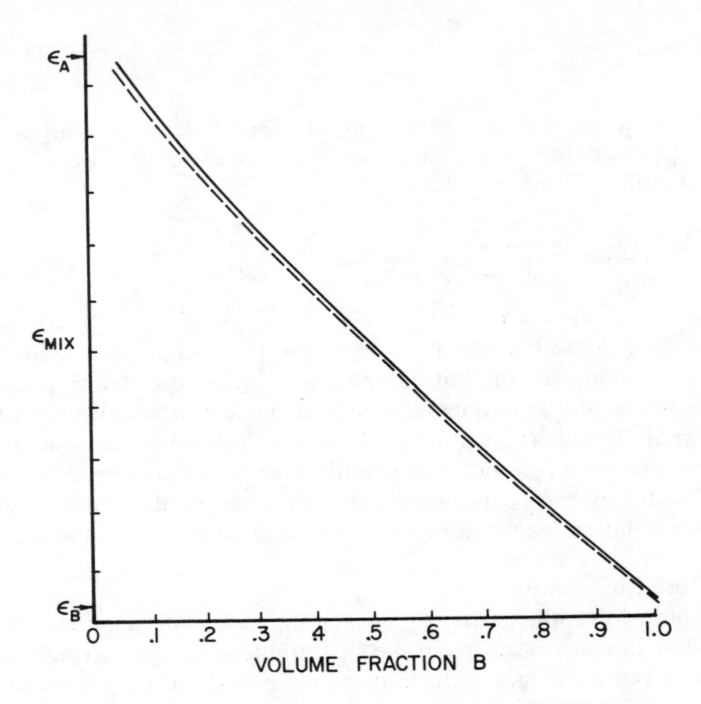

FIGURE 12-2. Relative agreement of experimental data (solid curve) for mixtures of two nonpolar liquids with Equation 12-42 (broken curve).

$$\overline{R} = f_R(p_d + \alpha\overline{R}) \tag{12-47}$$

Rearrangement of Equation 12-47 gives

$$\overline{R} = \frac{f_R p_d}{1 - f_R\alpha} \tag{12-48}$$

When the total effect of the reaction field, $\overline{R}_T$, is taken into account, the internal field is

$$\overline{E}_i = \overline{E}_d + \overline{R}_T \tag{12-49}$$

where $\overline{E}_d$ is that part of the field aligning, or directing, the dipoles. When the dipole moment is taken as the average of all orientations, Equation 12-48 becomes

$$< \overline{R}_T > = \frac{f_R}{1 - f_R\alpha} P_d \tag{12-50}$$

Equation 12-20 now is used for P_d, and $\overline{E}_d$ is used instead of $\overline{E}$, so that

$$\overline{R}_T = \frac{f_R}{1 - f_R\alpha} \frac{p_d^2 \overline{E}_d}{3k_B T} \tag{12-51}$$

gives the total reaction field for a single particle. This is substituted into Equation 12-49. Then

$$\overline{E}_i = \overline{E}_d + \frac{f_R}{1 - f_R\alpha} \frac{p_d^2 \overline{E}_d}{3k_BT} = \overline{E}_d \left[1 + \frac{f_R}{1 - f_R\alpha} \frac{p_d^2}{3k_BT}\right] \tag{12-52}$$

$\overline{E}_d$ may be expressed to include the effect of the reaction field as

$$\overline{E}_d = \overline{E}_c + f_R\alpha\overline{E}_d \tag{12-53}$$

where $\overline{E}_c$ is the field inside the hollow, spherical volume and is known as the cavity field. Equation 12-53 may be rewritten as

$$\overline{E}_d = \frac{1}{1 - f_R\alpha} \overline{E}_c \tag{12-54}$$

It can be shown[3] that the potential within the spherical cavity is

$$V_c = \frac{3\epsilon}{2\epsilon + \epsilon_1} \overline{E}_z$$

where ε and ε_1 are the dielectric constants of the surrounding material and the cavity, respectively; $\overline{E}_z$ is the uniform potential in z direction. The cavity being empty, except for the dipole at its center, has $\varepsilon_1 = 1$. Thus the cavity field is

$$\overline{E}_c = \frac{3}{2\epsilon + 1} \overline{E} \tag{12-55}$$

Equation 12-55 is always smaller than the Lorentz field (Equation 12-26).

Equation 12-55 is substituted into Equation 12-54, giving

$$\overline{E}_d = \frac{1}{1 - f_R\alpha} \frac{3}{2\epsilon + 1} \overline{E} \tag{12-56}$$

When Equation 12-56 is substituted into Equation 12-52, the internal field is

$$\overline{E}_i = \left[\frac{1}{1 - f_R\alpha} \frac{3\epsilon}{2\epsilon + 1}\right] \left[1 + \frac{f_R}{1 - f_R\alpha} \frac{p_d^2}{3k_BT}\right] \overline{E} \tag{12-57}$$

Use is now made of Equation 12-28a in the form employed by Onsager:

$$\frac{\overline{E}}{4\pi} (\epsilon - 1) = N \left[\alpha\overline{E}_i + \frac{p_d^2}{3k_BT} \overline{E}_d\right] \tag{12-58}$$

A comparison of Equations 12-28a and 12-58 shows the greater complexity of the Onsager relationship compared to that of Debye. In addition, a simplifyng substitution such as that which was made to obtain Equation 12-28b is not made in the Onsager approach. Instead, Equations 12-57 and 12-54 are used in Equation 12-58. When these substitutions are made, it is noted that $3\varepsilon/(2\varepsilon + 1)$ is a common factor and that $\overline{E}$ vanishes, it is found that

$$\frac{(\epsilon - 1)(2\epsilon + 1)}{12\pi\epsilon} = \frac{N}{1 - f_R\alpha} \left\{\alpha \left[1 + \frac{f_R}{1 - f_R\alpha} \frac{p_d^2}{3k_BT}\right] + \frac{p_d^2}{3k_BT}\right\} \tag{12-59}$$

Equation 12-59 is simplified as follows:

$$\frac{(\epsilon - 1)(2\epsilon + 1)}{12\pi\epsilon} = \frac{N}{1 - f_R\alpha}\left\{\alpha\left[\frac{(1 - f_R\alpha)\,3k_BT + f_Rp_d^2}{(1 - f_R\alpha)\,3k_BT}\right] + \frac{(1 - f_R\alpha)\,p_d^2}{(1 - f_R\alpha)\,3k_BT}\right\}$$

The indicated operations in the numerator are carried out using the common denominator. This results in

$$\frac{(\epsilon - 1)(2\epsilon + 1)}{12\pi\epsilon} = \frac{N}{1 - f_R\alpha}\left[\frac{3k_BT\alpha - 3k_BTf_R\alpha^2 + p_d^2}{(1 - f_R\alpha)\,3k_BT}\right]$$

$$\frac{(\epsilon - 1)(2\epsilon + 1)}{12\pi\epsilon} = \frac{N}{1 - f_R\alpha}\left[\frac{3k_BT\alpha\,(1 - f_R\alpha)}{(1 - f_R\alpha)\,3k_BT} + \frac{p_d^2}{3k_BT\,(1 - f_R\alpha)}\right]$$

$$\frac{(\epsilon - 1)(2\epsilon + 1)}{12\pi\epsilon} = \frac{N}{1 - f_R\alpha}\left[\alpha + \frac{p_d^2}{3k_BT\,(1 - f_R\alpha)}\right] \tag{12-60}$$

An expression for $1/(1 - f_R\alpha)$ may be obtained starting with Equation 12-46 to further clarify Equation 12-60:

$$f_Rr^3 = \frac{2(\epsilon - 1)}{2\epsilon + 1}\ ;\ r^3 = \frac{1}{f_R}\,\frac{2(\epsilon - 1)}{2\epsilon + 1} \tag{12-61}$$

Equation 12-61 now is substituted into Equation 12-44 to give

$$\frac{\epsilon_\infty - 1}{\epsilon_\infty + 2} = f_R\alpha\,\frac{2\epsilon + 1}{2(\epsilon - 1)}$$

This is rearranged as

$$\frac{\epsilon_\infty - 1}{\epsilon_\infty + 2}\,\frac{2(\epsilon - 1)}{2\epsilon + 1} = f_R\alpha \tag{12-62}$$

and then rewritten in the form

$$1 - \frac{\epsilon_\infty - 1}{\epsilon_\infty - 2}\,\frac{2(\epsilon - 1)}{2\epsilon + 1} = 1 - f_R\alpha \tag{12-63}$$

The reciprocal of this equation is

$$\frac{1}{1 - \dfrac{\epsilon_\infty - 1}{\epsilon_\infty + 2}\,\dfrac{2(\epsilon - 1)}{2\epsilon + 1}} = \frac{1}{1 - f_R\alpha} \tag{12-64}$$

The denominator of the fraction on the left side of Equation 12-64 is rewritten in terms of a common denominator $(\varepsilon_\infty + 2)(2\varepsilon + 1)$ and after simplification it is found that

$$\frac{(\epsilon_\infty + 2)\ (2\epsilon + 1)}{3\epsilon_\infty + 6\epsilon} = \frac{(\epsilon_\infty + 2)\ (2\epsilon + 1)}{3(\epsilon_\infty + 2\epsilon)} = \frac{1}{1 - f_R\alpha} \tag{12-65}$$

Equation 12-65 is substituted into Equation 12-60 to eliminate $1/(1 - f_R\alpha)$. Thus,

$$\frac{(\epsilon - 1)\ (2\epsilon + 1)}{12\pi\epsilon} = N \frac{(\epsilon_\infty + 2)\ (2\epsilon + 1)}{3(\epsilon_\infty + 2\epsilon)} \left[\alpha + \frac{p_d^2}{3k_BT} \frac{(\epsilon_\infty + 2)\ (2\epsilon + 1)}{3(\epsilon_\infty + 2\epsilon)}\right]$$

which reduces to

$$\frac{\epsilon - 1}{4\pi\epsilon} = N \frac{\epsilon_\infty + 2}{\epsilon_\infty + 2\epsilon} \left[\alpha + \frac{p_d^2}{3k_BT} \frac{(\epsilon_\infty + 2)\ (2\epsilon + 1)}{3(\epsilon_\infty + 2\epsilon)}\right] \tag{12-66}$$

The polarizability in Equation 12-66 is eliminated by the use of Equation 12-41 in the high-frequency range. This becomes

$$\alpha_e = \frac{3}{4\pi N} \frac{\epsilon_\infty - 1}{\epsilon_\infty + 2}$$

This is substituted into Equation 12-66 and employed as follows:

$$\frac{\epsilon - 1}{4\pi\epsilon} = N \frac{\epsilon_\infty + 2}{\epsilon_\infty + 2\epsilon} \left[\frac{3}{4\pi N} \frac{\epsilon_\infty - 1}{\epsilon_\infty + 2} + \frac{p_d^2}{3k_BT} \frac{(\epsilon_\infty + 2)\ (2\epsilon + 1)}{3(\epsilon_\infty + 2\epsilon)}\right]$$

or, multiplying the numerator and denominator of the last term in the brackets by $(\varepsilon_\infty + 2)$,

$$\frac{\epsilon - 1}{4\pi} = \frac{3\epsilon}{4\pi} \frac{\epsilon_\infty - 1}{\epsilon_\infty + 2} + \frac{Np_d^2}{3k_BT} \frac{(\epsilon_\infty + 2)^2\ (2\epsilon + 1)\epsilon}{3(\epsilon_\infty + 2\epsilon)^2} \tag{12-67}$$

Equation 12-67 can be reexpressed to give the Onsager equation in the following way: upon rearrangement,

$$\frac{\epsilon - 1}{4\pi} - \frac{Np_d^2}{9k_BT} \frac{(\epsilon_\infty + 2)^2\ (2\epsilon + 1)\epsilon}{(\epsilon_\infty + 2\epsilon)^2} = \frac{3\epsilon}{4\pi} \frac{\epsilon_\infty - 1}{\epsilon_\infty + 2\epsilon}$$

Or,

$$\frac{Np_d^2}{9k_BT} = \left[\frac{(\epsilon - 1)}{4\pi} - \frac{3\epsilon}{4\pi} \frac{\epsilon_\infty - 1}{\epsilon_\infty + 2}\right] \frac{(\epsilon_\infty + 2\epsilon)^2}{(\epsilon_\infty + 2)^2\ (2\epsilon + 1)\epsilon}$$

Then, performing the indicated multiplication,

$$p_d^2 = \frac{9k_BT}{4\pi N} \left[\frac{(\epsilon - 1)\ (\epsilon_\infty + 2\epsilon)^2}{(\epsilon_\infty + 2)^2\ (2\epsilon + 1)\epsilon} - \frac{3\epsilon(\epsilon_\infty - 1)\ (\epsilon_\infty + 2\epsilon)}{(\epsilon_\infty + 2)^2\ (2\epsilon + 1)\epsilon}\right]$$

The quantity within the brackets is expressed in terms of the common denominator, and factored to give

$$p_d^2 = \frac{9k_BT}{4\pi N} \left[\frac{(\epsilon_\infty + 2\epsilon)\ [(\epsilon - 1)\ (\epsilon_\infty + 2\epsilon) - 3\epsilon(\epsilon_\infty - 1)]}{(\epsilon_\infty + 2)^2\ (2\epsilon + 1)\epsilon}\right]$$

The numerator of the fraction is expanded and factored giving

$$p_d^2 = \frac{9k_BT}{4\pi N}\left[\frac{(\epsilon_\infty + 2\epsilon)\,[\epsilon(2\epsilon + 1) - \epsilon_\infty(2\epsilon + 1)]}{(\epsilon_\infty + 2)^2\,(2\epsilon + 1)\epsilon}\right]$$

Finally, since $(2\varepsilon + 1)$ vanishes, the expression for p_d^2 is found to be

$$p_d^2 = \frac{9k_BT}{4\pi N}\,\frac{(\epsilon_\infty + 2\epsilon)\,(\epsilon - \epsilon_\infty)}{(\epsilon_\infty + 2)^2\,\epsilon} \tag{12-68}$$

Equation 12-68 is the Onsager expression for the dipole moment of an atom or a molecule for a pure substance. It should be noted that the original derivation did not distinguish between $\overline{E}_d$ and $\overline{E}_i$ as was done in Equation 12-58.

Equation 12-68 permits the calculation of the dipole moments of pure, dipolar liquids or of concentrated liquid solutions; it explains the behavior of almost all polar molecules where the Debye equation is inaccurate and which require a correction factor as noted at the end of Section 12.4. The Onsager relationship does not hold where intermolecular associations (antiparallel molecular dipole alignments caused by short-range molecular forces) take place. Where no such interations are present in the liquid phase, the Onsager equation also gives reasonable results for molecules in the vapor phase, Equation 12-68 also is inaccurate when the molecules are nonspherical.

The selection of ε_∞ for use in Equation 12-68 is very important. It must be taken at frequencies which do not eliminate the atomic polarization contributions.

The Onsager equation also describes the dielectric behavior of an unusual class of substances. These materials include the hydrogen halides and some organic molecules which are very nearly spherical in shape. The molecules of both of these types of materials show an extremely high degree of ability to align themselves parallel to an external field when in the crystalline state. The values of p_d obtained for these substances in both the solid and liquid states are those which would be expected of the vapor phase. Crystalline materials of such globular molecules have been named ''rotator-phase'' solids because of their unusual ease of alignment.

The hydrogen halides show this behavior down to temperatures below about 200 K, where allotropic changes take place. Reversion to the more usual behavior occurs when the crystal lattices change. Examples of organic materials with spherical molecules include derivatives of such organic compounds as methane, camphane, and cyclohexane. The OH radicals of ice also rotate freely.

Rotator-phase molecules show the same degree of orientation mobility in the crystalline state as in the liquid state. Ordinarily it would be expected that the values of ε would be relatively high for the liquid phase and would show a large, virtually discontinuous decrease in ε, of about an order of magnitude, upon cooling through the freezing point (Figure 12-3a); the dipoles are not expected to be as free in the solid as they are in the liquid and P_d would ordinarily be smaller. In contrast to this expected behavior, some rotator-phase molecules may show different variations in ε as they pass through the freezing point. Some may show a continuing increase in ε, in the solid state, at temperatures below the liquid-solid transition temperature. Others have a discontinuous increase in ε upon solidification. Still others demonstrate continuously increasing values of ε as the temperature is lowered below the melting point, and then rapidly drop to relatively low values indicative of nonrotating behavior (Figure 12-3b). Some substances of this type show negative slopes above the melting point which are in approximate agreement with Equation 12-21.

12.5.3. Electronic and Ionic Polarization

Ionic solids, such as the alkali halides, provide an easily visualized example of a class of materials in which polarization effects result from both ionic and electronic contributions. These solids form regular lattices with high bonding energies (see Sections 10.6.3, 10.6.4, and Figure 10-3). The interpenetrating lattices of positive and negative ions tend to move with respect to each other in the presence of an applied electric field. Displacements of this kind are responsible for the ionic polarization.

The high bonding strengths of this class of compounds tend to limit the amount of such displacements and to restrict other ionic motions almost entirely to oscillations of the ions about their equilibrium positions (see Section 3.10 and Chapter 4, Volume I). These harmonic oscillations are coupled throughout both sublattices and determine the vibrational modes which can absorb energy (Sections 3.10 and 4.1.4). Thus, the ranges of frequencies to which these compounds can respond are directly related to this.

Neither of the ions of a solid belonging to this class have permanent dipoles. This, along with the strong bonding, eliminates the possibility of any orientation polarization. Even when such ions have nonsymmetrical charge distributions, they will not be aligned in definite orientations.

Electronic polarization is induced by the displacement of the electron cloud about each ion by the oscillating, applied electric field as described in Section 12.2. Both electronic and ionic polarizations contribute to the dielectric properties over a range of frequencies. However, a frequency is reached at which the electronic charge distributions around the ions can comply with the oscillations of the applied fields, but the ionic displacements cannot. The inertial effects of the ionic displacements prevent the ions from responding fast enough to react to these high frequencies. They, therefore, are unable to contribute to the polarization when this condition prevails. The frequency at which the ions are unable to respond occurs when the frequency of the applied field lies in the infrared range (Figure 12-7). The dielectric properties of this class of solids then become a function only of the electronic polarization when this situation occurs.

In an ionic crystal, such as an alkali halide, the ionic displacement is determined by equating the forces involved:

$$Kx = e\bar{E}_{loc}; \; x = e\bar{E}_{loc}/K \qquad (12\text{-}69)$$

where x is the displacement and K is a short-range interionic force constant. The ionic portion of the polarization, based on Equation 12-12, is

$$P_i = Nex \qquad (12\text{-}70)$$

Or, using Equation 12-69,

$$P_i = Ne^2\bar{E}_{loc}/K \qquad (12\text{-}71)$$

Thus, the polarizability of an ion in a static field is

$$\alpha_i = e^2/K \qquad (12\text{-}72)$$

Use now is made of Equation 12-38 in a more simplified form:

$$\frac{\epsilon - 1}{\epsilon + 2} = (4\pi/3)N\alpha = P \qquad (12\text{-}73)$$

Applying Equation 12-72 in Equation 12-73 gives

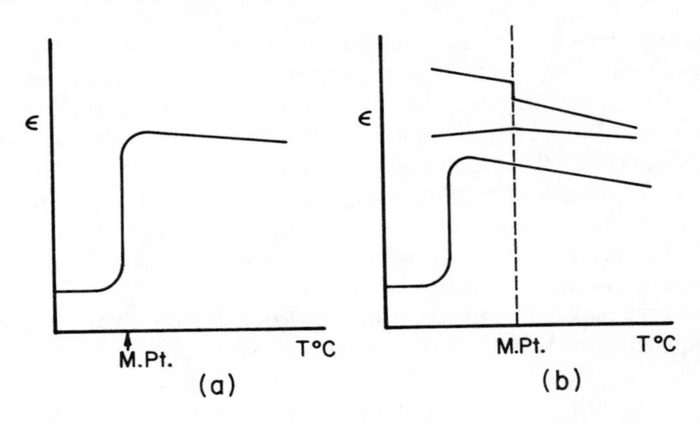

FIGURE 12-3. Schematic diagrams of (a) normal behavior of polar molecules and (b) behaviors of some "rotator-phase" molecules. Note the contrast between their properties at and below their melting points.

$$P_i = (4\pi/3)Ne^2/K \qquad (12\text{-}74)$$

The electronic portion is given, based upon Equation 12-38, as

$$P_e = (4\pi N/3)\,[\alpha(+)+\alpha(-)] \qquad (12\text{-}75)$$

where $\alpha_e = \alpha(+) + \alpha(-)$, the sum of the electronic polarizabilities of the positive and negative ions.

Equation 12-73 is reexpressed so that Equations 12-74 and 12-75 may be taken into account. This is rewritten as

$$\epsilon - 1 = P(\epsilon + 2) = P\epsilon + 2P$$

and rearranged as

$$\epsilon(1 - P) = 1 + 2P$$

This gives the dielectric constant as

$$\epsilon = \frac{1 + 2P}{1 - P} \qquad (12\text{-}76)$$

where

$$P = P_i + P_e \qquad (12\text{-}77)$$

and P_i and P_e are given by Equations 12-74 and 12-75, respectively.

Thus, for ionic solids, using Equation 12-77 in Equation 12-76,

$$\epsilon = \frac{1 + 2(P_i + P_e)}{1 - (P_i + P_e)} \qquad (12\text{-}78)$$

At high frequencies the ionic inertia is too great to permit the ions to respond so that the high-frequency dielectric constant may be obtained from Equation 12-78 as

$$\epsilon_\infty = \frac{1 + 2P_e}{1 - P_e} \qquad (12\text{-}79)$$

or, rearranging,

$$P_e = \frac{\epsilon_\infty - 1}{\epsilon_\infty + 2} \tag{12-79a}$$

Equations 12-78 and 12-79 also are useful in relating the optical properties of dielectrics to the longitudinal and transverse frequencies of their ions (see Equations 12-90 and 12-91).

The classical approach to obtain the motion of an oscillating particle (Equation 3-50, Volume I) can be used to obtain the frequency of phonos with long wave-lengths in the absence of an applied field. This is developed for oscillations perpendicular to the local field starting with

$$m\frac{d^2x}{dt^2} + Kx = e\bar{E}_{loc} = e\frac{4\pi}{3}P \tag{12-80}$$

where Equation 12-25 for $\bar{E}_{loc}$ is used, noting that no applied field is present.

When a molecule rather than a single particle is under consideration, it is necessary to use its reduced mass in Equation 12-80 and the equations which follow. The reason for this is that the components of the molecule oscillate about the center of mass of the molecule. This must be taken into account. Consider the molecule AB. The distance between atoms A and B at a given time, in one dimension, is the difference between their positions: $x_A - x_B$. If their equilibrium distance is r_o, then the change in their positions is

$$x = (x_A - x_B) - r_o \quad \text{and} \quad \frac{d^2x}{dt^2} = \frac{d^2x_A}{dt^2} - \frac{d^2x_B}{dt^2}$$

When $(x_A - x_B) > r_o$, A and B are farther apart than when at equilibrium; they are closer together when $(x_A - x_B) < r_o$. The use of Equation 3-50a for each atom gives

$$m_A\frac{d^2x_A}{dt^2} = -Kx$$

and

$$m_B\frac{d^2x_B}{dt^2} = +Kx$$

Now multiplying each equation by the mass of the other atom and taking the difference between them:

$$m_Am_B\frac{d^2x_A}{dt^2} - m_Am_B\frac{d^2x_B}{dt^2} = -m_BKx - m_AKx$$

This is rewritten as

$$\frac{m_Am_B}{m_A + m_B}\frac{d^2}{dt^2}(x_A - x_B) = -Kx$$

The fraction involving the masses is the reduced mass, m_R. And, noting the second derivative obtained earlier, the equation may be expressed as

$$m_R\frac{d^2x}{dt^2} = -Kx$$

This permits the treatment of the molecule as a single oscillator.

A more detailed expression is required for the factor P in order to obtain the desired solution to Equation 12-80. The total polarization is given by $P = P_i + P_e$. P_i is given by Equation 12-70. An expression for P_e is found starting with Equation 12-17 as applied to electronic polarization:

$$P_e = N\alpha_e \overline{E}_{loc}$$

Now, using Equation 12-80 for $\overline{E}_{loc}$ gives

$$P_e = N\alpha_e \frac{4}{3} \pi P$$

It will be recognized that $(4\pi N/3)\alpha_e$ is the same as Equation 12-75 for P_e. Thus,

$$P_e = P_e P$$

This relationship along with Equation 12-70 now is used to obtain the total polarization

$$P = P_i + P_e = Nex + P_e P$$

or,

$$P = \frac{Nex}{1 - P_e} \qquad (12\text{-}81)$$

This enables Equation 12-80 to be rewritten as

$$m \frac{d^2 x}{dt^2} = -Kx + e \frac{4\pi}{3} \frac{Nex}{1 - P_e}$$

and simplified by the use of Equation 12-74 to give

$$m \frac{d^2 x}{dt^2} = -Kx + \frac{KP_i x}{1 - P_e}$$

By the same means used to obtain Equation 3-55 (Volume I), and recalling that $\nu = 2\pi\upsilon$, $d^2x/dt^2 = -\omega_T{}^2 x$ and Equation 12-82 becomes

$$-m\omega_T{}^2 x = -Kx + \frac{KP_i x}{1 - P_e}$$

which simplifies to

$$m\omega_T{}^2 = K \left[1 - \frac{P_i}{1 - P_e} \right]$$

Equation 12-83 is useful for applications where the frequencies are in the optical range. When the transverse frequencies are small, the short-range interionic force constant, K, can be estimated. This approach is useful in the explanation of the effects of lattice transitions upon the properties of ferroelectric crystals (see Section 12.12.1 and Equation 12-244).

The case for oscillations parallel to the local field, in the absence of an external field, is developed starting with Equation 12-24 as

$$\bar{E}_{loc} = D - \frac{8\pi}{3} P = -\frac{8\pi}{3} P \qquad (12\text{-}84)$$

since D = 0 in this situation. In the same manner as for Equation 12-80, the equation of motion of the particle is, making use of Equation 12-84,

$$m \frac{d^2 x}{dt^2} + Kx = e\bar{E}_{loc} = -e \frac{8\pi}{3} P \qquad (12\text{-}85)$$

The total polarization is obtained from Equation 12-77, and noting that the coefficient of P in Equation 12-85 is twice that of Equation 12-80,

$$P = Nex - 2P_eP$$

or,

$$P = \frac{Nex}{1 + 2P_e} \qquad (12\text{-}86)$$

Following the same procedure as before, Equation 12-86 is substituted into Equation 12-85 to give

$$m \frac{d^2 x}{dt^2} + Kx = -e \frac{8\pi}{3} \frac{Nex}{1 + 2P_e} \qquad (12\text{-}87)$$

Again using Equation 12-74,

$$m \frac{d^2 x}{dt^2} + Kx = -\frac{2KP_ix}{1 + 2P_e}$$

Then, using the same approach as that used to obtain Equation 12-83,

$$-m\omega_L{}^2 x = -Kx - \frac{2KP_ix}{1 + 2P_e}$$

Since x vanishes, this reduces to

$$m\omega_L{}^2 = K \left[1 + \frac{2P_i}{1 + 2P_e} \right] \qquad (12\text{-}88)$$

Expressions now have been obtained involving both ω_T and ω_L in terms of their ionic and electronic polarizations (Equations 12-83 and 12-88, respectively). The ratio of these two equations is taken, noting that m and K vanish, giving

$$\frac{\omega_T{}^2}{\omega_L{}^2} = \frac{1 - \dfrac{P_i}{1 - P_e}}{1 + \dfrac{2P_i}{1 + 2P_e}} = \frac{1 + 2P_e}{1 - P_e} \cdot \frac{1 - (P_e + P_i)}{1 + 2(P_e + P_i)} \qquad (12\text{-}89)$$

The ratio of Equation 12-79 to Equation 12-78 is obtained as

$$\frac{\epsilon_\infty}{\epsilon} = \frac{\dfrac{1 + 2P_e}{1 - P_e}}{\dfrac{1 + 2(P_i + P_e)}{1 - (P_i + P_e)}} = \frac{1 + 2P_e}{1 - P_e} \cdot \frac{1 - (P_e + P_i)}{1 + 2(P_e + P_i)} \qquad (12\text{-}90)$$

The ratios given by Equations 12-89 and 12-90 are equal. This result gives the Lyddane-Sachs-Teller equation:

$$\frac{\omega_T^2}{\omega_L^2} = \frac{\epsilon_\infty}{\epsilon} \qquad (12\text{-}91)$$

This relates the dielectric constants and the transverse and longitudinal frequencies of the ions in the lattice by means of their ionic and electronic polarizations. In addition, it provides insight into the properties of ferroelectric materials which undergo phase changes.

The derivation of Equation 12-91 implicitly included the assumption that any interaction between the two sublattices is nonexistent in the absence of an external field. However, this neglects the varying electric dipole moments (fluctuations) which are always present in any molecule, ionic, or atomic particle. Even though the average, dipole moments are zero, the instantaneous dipole interactions lead to a nonzero force which could have been included in Equation 12-69 for greater accuracy. This force results from the interaction energy and is known as the dispersion energy or the London-Van der Waals energy. Data for Equation 12-91 are given in Table 12-4. Note that most of the differences are less than 10%. In summary, Equation 12-91, and Equations 12-78, 12-79, 12-83, and 12-88, from which it was derived, related the dielectric constants to the ionic frequencies, are helpful in estimating the interionic force constant, and assist in the explanation of the effects of changes in crystal structure upon optical and ferroelectric properties.

Some alkali halide crystals doped with divalent ions show increased ionic polarizability. A divalent ion in the crystal requires that an alkali ion be absent from the lattice in order to preserve electrical neutrality; it forms a defect lattice (see Sections 10.6.7, 10.6.9, and 12.9.1.3). The vacancy thus formed can move by the diffusion of an alkali ion into the original vacant lattice site. The vacancy moves to the site last occupied by the alkali ion (see Sections 12.9.1.2 and 12.9.1.3.). The motion of the vacancy changes, or rotates, the dipole of the original divalent ion-hole pair. This mechanism can give rise to enhanced values of P_i.

Nonpolar liquids have dielectric properties which may be described by a combination of Equations 12-29 and 12-38 as

$$P_m = \frac{\epsilon - 1}{\epsilon + 2}\frac{M}{\rho} = \frac{4\pi}{3}N\alpha = \frac{4\pi}{3}N(\alpha_e + \alpha_a) \qquad (12\text{-}92)$$

The effects of temperature are taken into account by the density of the liquid. Experimental work, using high frequencies, shows that $n^2 \simeq \alpha_e$. Thus, the values of α_a for these liquids are small. This is true for such liquids as bromine, carbon dioxide, carbon tetrachloride, and benzene (see also Table 12-3).

Nonpolar gases are well described by the conversion of a form of the Debye equation (Equation 12-29) into the Lorenz-Lorentz equivalent by means of Equation 12-39:

$$\frac{n^2 - 1}{n^2 + 2}\frac{M}{\rho} = \frac{4\pi}{3}N\alpha \qquad (12\text{-}93)$$

Table 12-4
STATIC AND HIGH-FREQUENCY DIELECTRIC CONSTANTS FOR SOME ALKALI HALIDES

Compound	ε	ε_∞	$\varepsilon_\infty/\varepsilon$	ω_T^2/ω_L^2 [a]
LiF	9.27	1.92	0.21	0.23
LiCl	11.05	2.75	0.25	0.23
LiBr	12.1	3.16	0.26	0.24
NaF	6.0	1.74	0.29	0.33
NaCl	5.62	2.62	0.44	0.41
KF	6.05	1.85	0.31	0.35
KCl	4.68	2.13	0.46	0.46
KI	4.94	2.69	0.55	0.56

[a] Calculated from data in Reference 2 (p. 156).

Selected from Dekker, A. J., *Solid State Physics*, Prentice-Hall, Englewood Cliffs, N.J., 1957, 145.

Data obtained for this class of substances at optical frequencies are slightly smaller than those obtained at lower frequencies; α_a makes no contribution. The molar polarization is not a function of pressure and temperature in systems at moderate pressures and temperatures. The Clausius-Mossotti "catastrophe" should occur when $\varrho = 3M/4\pi N\alpha$; here $\varepsilon - 1 = \varepsilon + 2$ and ε would have to be suitably large for this case to be meaningful. At normal pressures, these gases have ε of the order of unity.

Actually P_m for these gases increases slightly with pressure until their densities are about 8 mol/ℓ; after this P_m decreases. As examples, argon shows a maximum increase of about 0.7% at a pressure of about 200 atm; carbon dioxide shows an increase of about 2% at its maximum. These changes are explained by molecular interactions and fluctuations which increase P_m slightly. When the pressure becomes sufficiently high, the molecular repulsion diminishes the polarizability of a gas.

As would be expected, the molecular polarization of a given, nonpolar substance is higher in the gaseous state than in the liquid phase. The greater degree of intermolecular interaction in the liquid state almost completely represses any molecular fluctuations.

12.5.4. Electronic, Atomic, and Orientation Polarization

The class of substances whose molecules possess permanent dipoles derive their polarizations from all three sources. They may be described by combining Equations 12-31, 12-29, 12-38, and 12-28, respectively, as

$$P = P_e + P_a + P_d = \frac{\epsilon - 1}{\epsilon + 2}\frac{M}{\rho} = \frac{4\pi}{3} N \left[\alpha_e + \alpha_a + \frac{p_d^2}{3k_B T}\right] \tag{12-94}$$

This modification of the Clausius-Mossotti equation (Equation 12-38 is essentially the same as Equation 12-28b, the Debye equation.

As noted in Section 12.4, Equation 12-94 holds for gases at normal pressures. A plot of the total polarization vs. 1/T provides important information. The P-intercept ($1/T \rightarrow 0$) is $(4\pi N/3)(\alpha_e + \alpha_a)$ and the slope is $(4\pi N p_d^2/(9k_B)$. This permits the direct calculation: $p_d = 1.28 \times 10^{-20} [dP/d(1/T)]^{1/2}$ esu-cm.

Since no provision was made in the derivations of Equations 12-28b or 12-38 for fluctuations in the induced moments, corrections must be made where this factor affects the polarization. One such correction is given in Section 12.4. Several other corrections have been made for this effect.

The polarizability of gases is affected by pressure, as noted in the previous section. When the pressure becomes sufficiently high, repulsion between the gas molecules becomes important and causes smaller polarizations than predicted by Equation 12-94. At pressures lower than this, molecular attraction can cause higher polarizations.

Equation 12-94 cannot provide accurate results for substances in which antiparallel molecular associations take place. The greater degree of inter-molecular interactions in liquids, as compared to gases, almost completely eliminates this factor.

The Onsager equation (Equation 12-68) gives values for the dipole moments of gas molecules which are in good agreement for dilute liquid solutions in which molecular association does not take place. This equation indicates that dp_d/dT is independent of temperature. Experimental data show that this is not exactly the case. The derivation as given in Section 12.5.2 may be improved by basing it upon an ellipsoidal cavity with a molecular volume ($4\pi Nabc = 1$) which is the same as that of the hollow sphere. The complexity of this model limits its application.

Hydrogen-bonded liquids, such as alcohols and water, show virtually no change in P_d, but P_a is increased by the external field because the hydrogen ion is a proton with a single, positive charge. Such molecules appear to array themselves in chain-like structures (see Figure 12-4). Others form antiparallel complexes. P_a increases in the first situation, but P_d decreases in the latter case because some of the dipoles cancel each other. (Also see the comments upon the assumptions and limitations of the Onsager equation given in Section 12.5.2.)

Some of the polymers with polar structures, such as those based upon vinyl chloride, vinyl acetate, acrylates, and methylacrylates, frequently have larger values of ε than in their monomer liquid states. In such cases a distinct portion of the molecule is responsible for the orientation polarization. These molecules are joined to the chains and the dipolar portions can change their orientations only when corresponding portions of the molecular chains permit their motion. Variation in the positions of the dipoles along the chains are reflected in varying degrees of polarization. Some examples of dipolar arrays are given in Figure 12-4.

12.6. POLARIZATION IN ALTERNATING FIELDS

The effects of the frequency of an external electric field upon the polarizability and polarization of a substance have been mentioned in previous sections. This factor will be examined in greater detail here.

First consider the elastic displacement of the electron cloud about an atom or an ion, as described in Section 12.2, in a static field. The restoring force is obtained from Equation 12-14 which is of the form $F = Kx$, where K is $(Ze)^2/r^3$. The natural frequency of electrons treated as harmonic oscillators is obtained from Equation 3-55 (Volume II), recalling that $\omega_o = 2\pi\upsilon$ and substituting for K, as

$$\omega_o^{\,2} = \frac{K}{m} = \frac{(Ze)^2}{mr^3} \qquad (12\text{-}95)$$

where r is the most probable radius of the electron "cloud". Neglecting local field effects, the restoring force is, from Equation 12-95,

$$F = Kx = m\omega_o^{\,2}x \qquad (12\text{-}96)$$

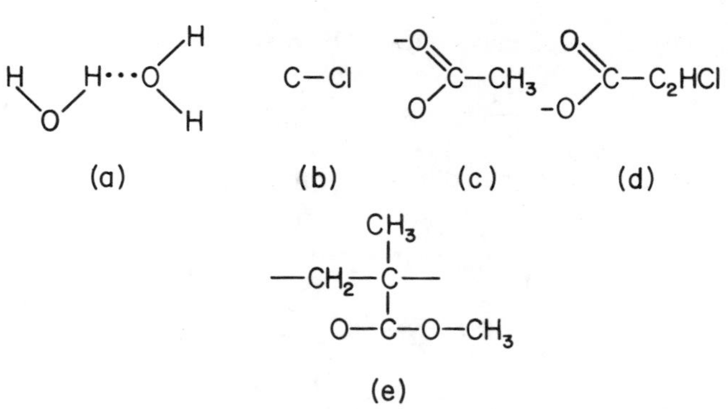

FIGURE 12-4. (a) Hydrogen bond in water. Dipolar portions of molecules: (b) polyvinyl chloride; (c) polyvinyl acetate; (d) polyvinyl chloracetate; and (e) polymethyl acrylate.

In a constant field $\overline{E}$

$$F = \overline{E}e = m\omega_o^2 x \tag{12-97}$$

and

$$\frac{x}{\overline{E}} = \frac{e}{m\omega_o^2}; \quad x = \frac{e\overline{E}}{m\omega_o^2} \tag{12-98}$$

Multiplying both sides of Equation 12-98 by e gives

$$\frac{ex}{\overline{E}} = \frac{e^2}{m\omega_o^2} \tag{12-99}$$

It will be recalled from Equation 12-12 that p = ex, and, from Equation 12-13 that $p/\overline{E} = \alpha_e$, so that Equation 12-99 gives the electronic polarizability in a constant field as

$$\frac{p}{\overline{E}} = \alpha_e = \frac{e^2}{m\omega_o^2} \tag{12-100}$$

In an electric field which varies periodically

$$\overline{E} = \overline{E}_o \exp(i\omega t) \tag{12-101}$$

where $\overline{E}_o$ is the maximum value of the electric field and t is time. So, for free electrons

$$F = ma = m\frac{d^2x}{dt^2} = e\overline{E}_o \exp(i\omega t) \tag{12-102}$$

Equation 12-102 describes the behavior of valence electrons in a metal and other cases in which electrons may be excited to nearly free behavior. However, in dielectric materials the electrons are tightly bound in the ion cores so that Equation 12-97 must be included in Equation 12-102 to account for this:

$$m\frac{d^2x}{dt^2} + m\omega_o^2 x = e\overline{E} \exp(i\omega t) \tag{12-103}$$

A solution to Equation 12-103 may be found by reexpressing it as

$$\frac{d^2x}{dt^2} + \omega^2 x = (e\bar{E}/m)\exp(i\omega t)$$

Or, in terms of operators, designated by the symbol D, this becomes

$$(D^2 + \omega^2)x = 0$$

so that

$$D^2 = -\omega^2;\ D = i\omega$$

Then,

$$x = x_o \exp(i\omega t) \tag{12-104}$$

The substitution of Equation 12-104 into Equation 12-103 results in

$$-m\omega^2 x_o \exp(i\omega t) + m\omega_o^{\,2} x_o \exp(i\omega t) = e\bar{E}_o \exp(i\omega t)$$

The exponential factors vanish and

$$mx_o(\omega_o^{\,2} - \omega^2) = e\bar{E}_o$$

Upon rearrangement the amplitude of the displacement is found to be

$$x_o = \frac{e\bar{E}_o}{m} \cdot \frac{1}{\omega_o^{\,2} - \omega^2} \tag{12-105}$$

This permits the calculation of the dipole moment per unit volume using Equations 12-17 and 12-105, as well as the polarizability (Equation 12-13)

$$P = Np = Nex = \frac{Ne^2\bar{E}_o}{m(\omega_o^{\,2} - \omega^2)} \quad \text{where } \alpha_e = \frac{e^2}{m(\omega_o^{\,2} - \omega^2)} \tag{12-106}$$

The response of a dipole or an electron cloud to an electric field is not instantaneous, but lags behind the oscillations of the field. This can result from anharmonic thermal coupling (see Section 4.4 and Figures 4-20 and 4-22, Volume I). and/or from inertial effects which are known as damping (see Sections 12.7 and 12.8). It will be noted that Equation 12-103 contains no terms to account for these effects so that Equation 12-106 is valid only where little or no damping is present. Both Equations 12-105 and 12-106 contain a singularity at $\omega = \omega_o$. This is shown schematically in Figure 12-5.

Where ω is much less than ω_o, the polarizability approaches that in a constant electric field as given by Equation 12-100. This may be shown by the use of Equation 12-106 in the form

$$\alpha_e = \frac{P}{N\bar{E}} = \frac{e^2}{m(\omega_o^{\,2} - \omega^2)} \rightarrow \frac{e^2}{m\omega_o^{\,2}};\ \omega << \omega_o \tag{12-107}$$

When ω is greater than ω_o, the oscillating dipole cannot respond because the frequency is greater than its natural, or resonant, frequency. These responses have interesting effects upon the polarization (Equation 12-106 the dielectric constant (Equation 12-

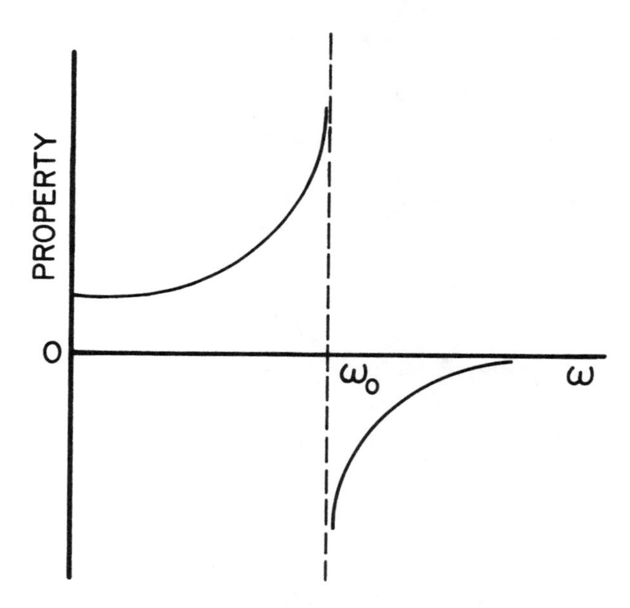

FIGURE 12-5. Schematic effect of frequency upon such properties as amplitude of displacement, x_o, polarization, P and polarizability, α_e (see Figure 12-13a).

79a), and upon the index of refraction of a dielectric material (Section 12.11). The polarization increases slowly, at first, as ω increases. The dielectric constant increases correspondingly. The material has a real index of refraction at frequencies at and below the ultraviolet, and is transparent. However, as $\omega \rightarrow \omega_o$ dispersion takes place; that is, the polarization increases rapidly and ε_∞ increases accordingly until the singularity is approached. In the range in which $\omega > \omega_o$, P, and ε_∞ are negative and the index of refraction is imaginary; total reflection takes place in this range of frequencies (see Section 12.11).

Implicit in Equation 12-103 is the assumption of instantaneous response of the electrons to the applied field. However, a time for response (damping) is involved and must be taken into account, since the electron response is not instantaneous. This effect is represented by means of the term Bmdx/dt which is included in Equation 12-103 in which B is the damping coefficient:

$$m \frac{d^2x}{dt^2} + Bm \frac{dx}{dt} + m\omega_o{}^2 x = e\bar{E} \exp(i\omega t) \qquad (12\text{-}108)$$

Equation 12-104 is used again and Equation 12-108 becomes, after noting that the exponential factor vanishes,

$$-\omega^2 x_o + Bi\omega x_o + \omega_o{}^2 x_o = \frac{e\bar{E}_o}{m} \qquad (12\text{-}109)$$

or

$$x_o[(\omega_o{}^2 - \omega^2) + Bi\omega] = \frac{e\bar{E}_o}{m}$$

and

$$x_o = \frac{e\bar{E}_o}{m} \frac{1}{(\omega_o^2 - \omega^2) + Bi\omega} \tag{12-110a}$$

so that

$$P = Nex = \frac{Ne^2\bar{E}_o}{m[(\omega_o^2 - \omega^2) + Bi\omega]} \tag{12-110b}$$

where, using Equations 12-12 and 12-13,

$$\alpha_e = \frac{e^2}{m[(\omega_o^2 - \omega^2) + Bi\omega]} \tag{12-110c}$$

It will be observed that for a static field $\omega = 0$ and Equation 12-110a reduces to

$$x_o = \frac{e\bar{E}_o}{m\omega_o^2} \tag{12-111}$$

which is the same as Equation 12-98. In addition, Equation 12-110c becomes

$$\alpha_e = \frac{e^2}{m\omega_o^2} \tag{12-112}$$

for the same condition, and is the same as Equation 12-100. When one electron is being considered, and e and m are its respective charge and mass, Equation 12-112 gives its effect upon thc polarizability. As shown from Equation 12-15, α_e is approximately 10^{-24} cm³. The use of this value in Equation 12-112 gives the natural frequency $\nu = \omega_o/2\pi$ of about 10^{15}/sec. The electronic polarizability can be approximated as being constant through the range of frequencies from below the microwave region up to the ultraviolet. When an ion or an atom is being considered, its natural frequency is found to be about 10^{13}/sec. This corresponds to the infrared range of frequencies.

The high-frequency dielectric constant may be obtained from equation 12-110 by starting with Equation 12-9 and rearranging it as

$$\bar{E}\epsilon_\infty = 4\pi P + \bar{E}$$

From which

$$\epsilon_\infty = \frac{4\pi P}{\bar{E}} + 1 \tag{12-113}$$

Then, using Equation 12-12 in the form P = Nex and substituting this into Equation 12-113 gives

$$\epsilon_\infty = \frac{4\pi Nex}{\bar{E}} + 1$$

Now, Equation 12-110a is used in its more general form to obtain

$$\epsilon_\infty = \frac{4\pi Ne}{\bar{E}} \frac{e\bar{E}}{m} \frac{1}{(\omega_o^2 - \omega^2) + Bi\omega^2} + 1$$

or

$$\epsilon_{\infty} = \frac{4\pi Ne^2}{m} \frac{1}{(\omega_o^2 - \omega^2) + Bi\omega^2} + 1 \quad (12\text{-}114)$$

Upon rationalization, Equation 12-114 becomes

$$\epsilon_{\infty} = \frac{4\pi Ne^2}{m} \left[\frac{(\omega_o^2 - \omega^2) - Bi\omega}{(\omega_o^2 - \omega^2)^2 + B^2\omega^2} \right] + 1 \quad (12\text{-}115)$$

This complex function can be separated into its real and imaginary parts by letting

$$\epsilon_{\infty} = \epsilon' - i\epsilon''$$

Thus the real part is

$$\epsilon' = \frac{4\pi Ne^2}{m} \frac{\omega_o^2 - \omega^2}{(\omega_o^2 - \omega^2)^2 + B^2\omega^2} + 1 \quad (12\text{-}116a)$$

and the imaginary part is

$$\epsilon'' = \frac{4\pi Ne^2}{m} \frac{B\omega}{(\omega_o^2 - \omega^2)^2 + B^2\omega^2} \quad (12\text{-}116b)$$

Equation 12-116a is responsible for the polarization of the dielectric material because it represents that portion of the dielectric constant which results from the oscillations being in phase with the field (see Equation 12-79a). ε'' in Equation 12-116b is proportional to the damping coefficient B, and, therefore, is responsible for the energy loss or the power loss (also see Section 12.7). The real and imaginary parts of Equation 12-116 are shown in Figure 12-6 at a resonant frequency.

The presence of B in the numerator of Equation 12-116b accounts for the damping and, consequently, for energy losses. Absorption would not take place if no damping occurred. This is called resonance absorption. The real part of the dielectric constant, ε', is a pronounced function of frequency in the ranges $\omega \rightarrow \omega_o$. Several resonant frequencies may be shown by many dielectric materials. The resultant curves for such materials reflect this and may contain several resonant absorptions of the kind shown in Figure 12-6. This behavior is shown schematically in Figure 12-7. Also see Figure 12-13.

The polarizability is greatest at those frequencies at which the dipoles can oscillate in response to the field. As the frequency is increased, a point is reached at which the dipoles can no longer respond to the alternating field and α_d is damped out. Here, the polarizability decreases and is the sum of the elastic displacements of atomic/ionic components plus the electronic polarizability. As the infrared range is approached the elastic displacements reach a resonant frequency and behave as shown in Figure 12-6a. Beyond this, the polarizability results from oscillations of the electron cloud until, in the ultraviolet range, another resonant frequency is reached. Here, as noted earlier in this section, dispersion takes place.

12.7. DIELECTRIC LOSS AND RELAXATION TIME

Previous discussions have shown that the dielectric polarization arises from electronic, atomic/ionic, and dipolar components:

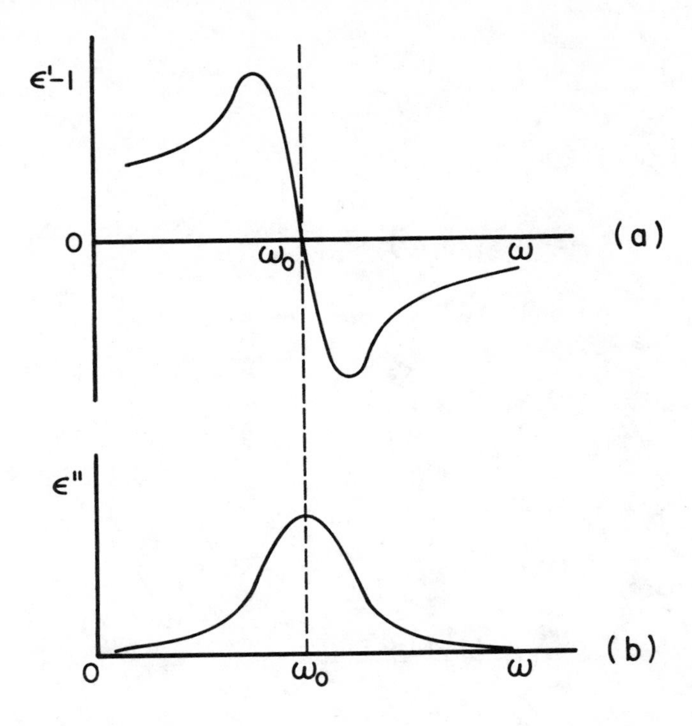

FIGURE 12-6. (a) Real part of the dielectric constant. (b) The imaginary part of the dielectric constant which is responsible for losses. Both are shown in the range in which ω approaches ω_o.

$$P = P_e + P_a + P_d \tag{12-31}$$

A given time is required for P to reach its maximum value. The behavior of P_d is of chief interest here since P_e and P_a reach their maximum values at rates which may be approximated as being instantaneous with respect to P_d. This behavior implies that the frequencies involved must be less than those corresponding to infrared frequencies (see Figure 12-7).

Let P_o be the maximum value of P_d. The degree of polarization may be expressed by

$$P_d = P_o(1 - e^{-t/\tau}) \tag{12-117}$$

in which t is the time and τ is the relaxation time. When the field is turned off, the decay of polarization will be proportional to $\exp(-t/\tau)$. Here, τ usually is defined as the time for the polarization to change (rise or fall) by an amount P_o/e. Upon differentiation of Equation 12-117

$$\frac{dP_d}{dt} = -P_o e^{-t/\tau}(-1/\tau) = \frac{1}{\tau} P_o e^{-t/\tau} . \tag{12-118}$$

Equation 12-117 may be expanded and rearranged as

$$P_o e^{-t/\tau} = P_o - P_d \tag{12-119}$$

Equation 12-119 is substituted into Equation 12-118:

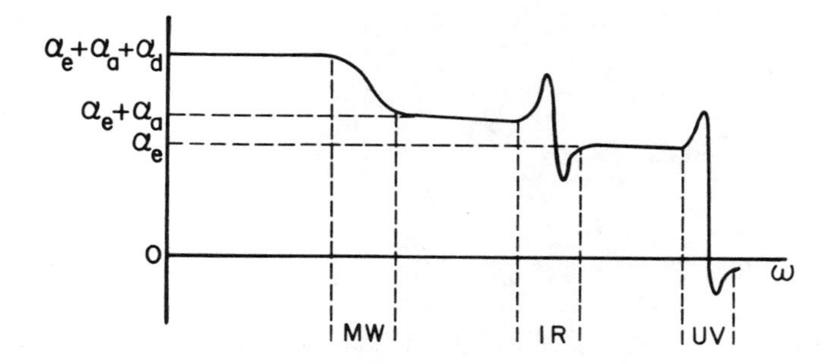

FIGURE 12-7. Real components of the polarizability of a dipolar substance showing the effects of the resonant frequencies in the microwave, infrared, and ultraviolet ranges (schematic).

$$\frac{dP_d}{dt} = \frac{1}{\tau}\,[P_o - P_d] \tag{12-120}$$

For an electric field which varies periodically, use is made of Equation 12-101:

$$\bar{E} = \bar{E}_o \exp(i\omega t) = \bar{E}(t) \tag{12-121}$$

In an alternating field P_o will be a function of time, P(t). Here, P(t) is that degree of polarization which corresponds to an instantaneous value of $\bar{E}$ given by Equation 12-121. So, for this condition Equation 12-120 becomes

$$\frac{dP_d}{dt} = \frac{1}{\tau}\,[P(t) - P_d] \tag{12-122}$$

Since P_e and P_a reach their maximum values very quickly, another dielectric constant, ε_p, can be defined, based upon Equation 12-9 as

$$P_e + P_a = \frac{\bar{E}}{4\pi}(\epsilon_p - 1) \tag{12-123}$$

The saturation value of P_d is found from Equation 12-31 by means of Equation 12-123:

$$P_o = P - (P_e + P_a) = \frac{\bar{E}}{4\pi}(\epsilon - 1) - \frac{\bar{E}}{4\pi}(\epsilon_p - 1)$$

or,

$$P_o = \frac{\bar{E}}{4\pi}(\epsilon - \epsilon_p) \tag{12-124}$$

Equation 12-124 is substituted into Equation 12-122 for P(t):

$$\frac{dP_d}{dt} = \frac{1}{\tau}\left[\frac{\bar{E}}{4\pi}(\epsilon - \epsilon_p) - P_d\right] \tag{12-125}$$

Now substituting Equation 12-121 into Equation 12-125,

$$\frac{dP_d}{dt} = \frac{1}{\tau}\left[\frac{\epsilon - \epsilon_p}{4\pi}\bar{E}_o e^{i\omega t} - P_d\right] \quad (12\text{-}126)$$

This may be rearranged as

$$\tau \frac{dP_d}{dt} + P_d = Ce^{i\omega t}; \; C = \frac{\epsilon - \epsilon_p}{4\pi}\bar{E}_o \quad (12\text{-}127)$$

The solution to Equation 12-127 is obtained from the sum of the natural response, P_n, and the steady-state response, P_s. The natural, or transient, response is given by the function

$$\tau \frac{dP_n}{dt} + P_n = 0 \quad (12\text{-}128)$$

Assume a trial solution

$$P_n = Ae^{\lambda t} \quad (12\text{-}129)$$

The substitution of Equation 12-129 into Equation 12-128 gives

$$\tau\lambda Ae^{\lambda t} + Ae^{\lambda t} = 0$$

The factor $A\exp(\lambda t)$ vanishes and

$$\tau\lambda + 1 = 0$$

Thus,

$$\lambda = -\frac{1}{\tau}$$

This gives the transient response as

$$P_n = Ae^{-t/\tau} \quad (12\text{-}130)$$

Assume, for the steady-state case, that

$$P_s = Be^{i\omega t} \quad (12\text{-}131)$$

Then, substituting for P_s in Equation 12-127,

$$\tau \frac{dP_s}{dt} + P_s = Ce^{i\omega t} \quad (12\text{-}132)$$

The use of Equation 12-131 in Equation 12-132 gives

$$i\omega\tau Be^{i\omega t} + Be^{i\omega t} = Ce^{i\omega t}$$

The exponential factors vanish and

$$B(i\omega\tau + 1) = C$$

or,

$$B = \frac{C}{1 + i\omega\tau}$$

and

$$P_s = \frac{C}{1 + i\omega\tau} e^{i\omega t} \qquad (12\text{-}133)$$

The complete solution to Equation 12-127 is given by the sum of Equations 12-130 and 12-133:

$$P(t) = Ae^{-t/\tau} + C \frac{e^{i\omega t}}{1 + i\omega\tau}$$

The expression for C is introduced from Equation 12-127 to obtain

$$P(t) = Ae^{-t/\tau} + \frac{\epsilon - \epsilon_p}{4\pi} \frac{\bar{E}_o e^{i\omega t}}{1 + i\omega\tau} \qquad (12\text{-}134)$$

which is the sum of transient and steady-state responses.

The steady-state response is of importance in obtaining the dielectric constant for frequencies lower than those discussed in Section 12.6. The frequencies under consideration here are less than those of the infrared range. Here the polarization is sufficiently rapid so that the transient response may be neglected in the following analysis. This is done by starting with Equation 12-9 in the form

$$4\pi P = \bar{E}(t)\, \epsilon\, (\omega) - \bar{E}(t)$$

and rewriting it as

$$\bar{E}(t)\, \epsilon\, (\omega) = 4\pi P + \bar{E}(t) \qquad (12\text{-}135)$$

The polarization now is given by $P = P_e + P_a + P(t)$. This is included in Equation 12-135 to obtain

$$\bar{E}(t)\, \epsilon\, (\omega) = 4\pi\, [P_e + P_a + P(t)] + \bar{E}(t) \qquad (12\text{-}136)$$

Equation 12-123 and the steady-state portion of Equation 12-134 are substituted into Equation 12-136 to get

$$\bar{E}(t)\, \epsilon\, (\omega) = 4\pi \left[\frac{\epsilon_p - 1}{4\pi} \bar{E}(t) + \frac{\epsilon - \epsilon_p}{4\pi} \frac{\bar{E}_o e^{i\omega t}}{1 + i\omega\tau} \right] + \bar{E}(t)$$

Now, noting that the factor 4π vanishes, and using Equation 12-121,

$$\bar{E}_o e^{i\omega t} \epsilon(\omega) = (\epsilon_p - 1)\, \bar{E}_o e^{i\omega t} + (\epsilon - \epsilon_p) \frac{\bar{E}_o e^{i\omega t}}{1 + i\omega\tau} + \bar{E}_o e^{i\omega t}$$

The factors $\bar{E}_o^{i\omega t}$ vanish leaving

$$\epsilon(\omega) = \epsilon_p - 1 + \frac{\epsilon - \epsilon_p}{1 + i\omega\tau} + 1 = \epsilon_p + \frac{\epsilon - \epsilon_p}{1 + i\omega\tau} \qquad (12\text{-}137)$$

The rationalization of the denominator of the fraction in Equation 12-137 gives

$$\epsilon(\omega) = \epsilon_p + (\epsilon - \epsilon_p) \frac{1 - i\omega\tau}{1 + \omega^2\tau^2} \tag{12-138}$$

The real and imaginary parts of Equation 12-138 are separated by letting

$$\epsilon(\omega) = \epsilon'(\omega) - i\epsilon''(\omega)$$

This results in

$$\epsilon'(\omega) = \epsilon_p + \frac{\epsilon - \epsilon_p}{1 + \omega^2\tau^2} \tag{12-139a}$$

and

$$\epsilon''(\omega) = (\epsilon - \epsilon_p) \frac{\omega\tau}{1 + \omega^2\tau^2} \tag{12-139b}$$

The dielectric power loss is proportional to Equation 12-139b (see Section 12.8). Equation 12-139b has a maximum at $\omega\tau = 1$. Then, the maximum losses of various dielectrics occur at frequencies which correspond to $1/\tau$. At frequencies much less than $1/\tau$, $\varepsilon'(\omega)$ approaches the value of the static dielectric constant, ε. In this low-frequency regime, the dipoles are able to oscillate and can contribute fully to the polarization; the losses are virtually zero. At frequencies greater than those corresponding to $1/\tau$ the dipoles cannot oscillate as quickly as the field. In this case $\varepsilon'(\omega)$ approaches ε_p. This behavior is shown in Figure 12-8 and explains the initial portion of Figure 12-7.

It should be noted that Equations 12-139a and 12-139b, often called the Debye equations, may be obtained from other models. In addition, it is frequently the case that experimental data can be explained only by a range of relaxation times instead of a single relaxation time. When this is the case, the Debye equations may be modified by a distribution function for the relaxation times.

12.8. DIELECTRIC ENERGY LOSSES

The dielectric constants derived for periodically varying fields given by Equations 12-114 and 12-137 are complex and are functions of time. Equation 12-11 relates the dielectric constant to the displacement and the polarization in a given field:

$$\bar{E}\epsilon = D = 4\pi P + \bar{E} \tag{12-11}$$

This means that P and D must have the same properties. However, P and D can lag behind $\bar{E}$ (recall the discussion leading to Equation 12-108).

If the alternating field is given by

$$\bar{E} = \bar{E}_o \cos \omega t \tag{12-140}$$

then, whre δ represents the angular difference in phase between $\bar{E}$ and D, or the phase angle, then

$$D = D_o \cos(\omega t - \delta) = D_o \cos \omega t \cos \delta + D_o \sin \omega t \sin \delta$$

or,

$$D = D_o \cos(\omega t - \delta) = D' \cos \omega t + D'' \sin \omega t \tag{12-141}$$

where

$$D' = D_o \cos \delta \quad \text{and} \quad D'' = D_o \sin \delta \tag{12-142}$$

A proportionality between D and $\overline{E}$ may be obtained from Equation 12-5 for most dielectric materials in static fields. However, the ratio of $D/\overline{E}$ usually is dependent upon frequency. This may be expressed most simply by using Equations 12-5 and 12-121

$$D = \epsilon(\omega)\overline{E} = \epsilon(\omega)\overline{E}_o e^{i\omega t} \tag{12-143}$$

keeping in mind that $\varepsilon(\omega)$ is frequency-dependent and complex. Two frequency-dependent dielectric constants may be defined, using Equation 12-142, as

$$\epsilon'(\omega) = \frac{D'}{\overline{E}_o} = \frac{D_o \cos \delta}{\overline{E}_o} \tag{12-144a}$$

and

$$\epsilon''(\omega) = \frac{D''}{\overline{E}_o} = \frac{D_o \sin \delta}{\overline{E}_o} \tag{12-144b}$$

These relationships can be used to give a complex dielectric constant of the form

$$\epsilon(\omega) = \epsilon'(\omega) - i\epsilon''(\omega) \tag{12-145}$$

Now, taking the ratio of Equation 12-144b to 12-144a, and noting that $D_o/\overline{E}_o$ vanishes,

$$\frac{\epsilon''(\omega)}{\epsilon'(\omega)} = \frac{\sin \delta}{\cos \delta} = \tan \delta \tag{12-146}$$

The phase angle is frequency-dependent because of $\varepsilon'(\omega)$ and $\varepsilon''(\omega)$.

Use was made in previous sections of the fact that dielectric power losses are proportional to the imaginary part of the dielectric constant. The basis for this relationship now will be derived. In a dielectric material the energy consumed by a unit volume in unit time is

$$U = \frac{\omega}{2\pi} \int_0^{2\pi/\omega} \frac{dq}{dt} \overline{E} dt \tag{12-147}$$

in which dq/dt is a current density. An expression for dq/dt is found using Equation 12-2 and the derivative of Equation 12-141:

$$\frac{dq}{dt} = \frac{1}{4\pi}\frac{dD}{dt} = \frac{\omega}{4\pi}(-D' \sin \omega t + D'' \cos \omega t) \tag{12-148}$$

Substituting Equations 12-148 and 12-140 into Equation 12-147 gives

$$U = \frac{\omega}{2\pi} \int_0^{2\pi/\omega} \frac{\omega}{4\pi} (-D' \sin \omega t + D'' \cos \omega t)\, \bar{E}_o \cos \omega t dt \qquad (12\text{-}149)$$

Simplifying,

$$U = \frac{\omega^2}{8\pi^2} \int_0^{2\pi/\omega} (-D' \sin \omega t \cos \omega t + D'' \cos^2 \omega t)\, \bar{E}_o dt$$

Upon integration

$$U = \frac{\omega^2 \bar{E}_o}{8\pi^2} \left[-\frac{D' \sin^2 \omega t}{2\omega} + \frac{D''}{\omega} \left(\frac{1}{2} \sin \omega t \cos \omega t + \frac{1}{2} \omega t \right) \right]_0^{2\pi/\omega}$$

Applying the limits gives

$$U = \frac{\omega \bar{E}_o}{8\pi^2} \left[0 + D'' \frac{\omega}{2} \frac{2\pi}{\omega} \right] = \frac{\omega \bar{E}_o}{8\pi} D''$$

Now, multiplying numerator and denominator by E_o and using Equation 12-144b for D'', and Equation 12-146 for $\varepsilon''(\omega)$,

$$U = \frac{\omega \bar{E}_o^{\,2}}{8\pi} \frac{D''}{\bar{E}_o} = \frac{\omega \bar{E}_o^{\,2}}{8\pi} \epsilon''(\omega) = \frac{\omega \bar{E}_o^{\,2}}{8\pi} \epsilon'(\omega) \tan \delta \qquad (12\text{-}150)$$

Thus, it has been shown that the energy losses are directly proportional to the imaginary part of the dielectric constant. It will be noted that the energy losses are proportional to $\sin \delta$ because $\varepsilon''(\omega)$ is a function of $\sin \delta$ (Equation 12-144b). It is for this reason that δ is called the loss angle and $\sin \delta$ the loss factor. Equation 12-146 for $\tan \delta$ often is incorrectly termed the loss factor. This is true only when δ is sufficiently small so that $\tan \delta \simeq \sin \delta$.

In applications of dielectric materials for use at high frequencies, materials with low losses are desirable. This means that such materials should have a minimum of dipolar polarization. Some applications make use of dielectric losses for domestic and industrial heating purposes. This method of heating, as in dielectric cookers, or ovens, has the advantage of heating the volume of the material quickly and uniformly.

12.9. CONDUCTION IN DIELECTRIC MATERIALS

All materials show some degree of electrical condictivity. The properties of metals and semiconductors are described in Chapters t (Volume I), 6 (Volume II), and 11. Most materials obey Ohm's law. However, as the strength of the electric field is increased, the conductivity of dielectrics becomes a function of the field strength. Solid materials of this kind may undergo lasting damage, or be destroyed, when subjected to sufficiently high fields.

Dielectric materials may have electronic or ionic components of their conductivities. Both mechanisms may be operative in some cases.

12.9.1. Ionic Conduction

Electrical conduction in ionic crystals can result from the movement of the ions in the lattice. This is called intrinsic conductivity and is important at high temperatures.

Conduction also may result from the presence of impurity ions in the lattice. This is called extrinsic conductivity and may occur at relatively low temperatures. In an analogous way, the conductivity observed in molecular crystals and polymers usually results from the extrinsic behavior caused by weakly bound impurity ions present in these materials.

The bonding energies of ionic crystals are high (Table 10-11). This indicates that an external field of the order of 10^6 to 10^8 V/cm is required for ionic conduction to take place in "perfect" crystals. However, fields of much lower strengths than these can cause a current to flow in ionic solids. This leads to the explanation of ionic conductivity in terms of the lattice defects present in these substances.

The two types of thermally induced lattice defects employed to explain this behavior are Frenkel and Schottky defects (Figure 12-9).

A Frenkel defect is caused by the motion of an ion from a lattice site to an interstitial position; two point defects are created: a vacancy and an interstitial. It is most probable that such interstitials are a result of a series of jumps rather than the jump of just one ion. A Schottky defect is the result of the migration of an ion from a lattice site to the surface; thus a vacancy is created. Since electrical neutrality must, on the average, be maintained both upon the surface and within the volume of the crystal, Schottky defects must be created in pairs of opposite sign. The concentration of Schottky defects is very much greater than that of Frenkel defects in ionic crystals. Ionic transport can occur by the jumping of adjacent ions into the lattice vacancies induced by both defect mechanisms. In a way analogous to that given for electrons and holes in Section 11.1.2, the behavior of the ions may be described by the motion of the vacancies. This consists of a series of ion jumps from lattice sites to adjacent, vacant lattice sites. The vacancies thus move in directions opposite to those of the ions.

12.9.1.1. Statistical Treatment of Intrinsic Conduction

An ion, and thus a vacancy, moves through the lattice by jumping over the potential barrier between one site and the next. All of these barriers are identical and each lattice site for a given species is the equivalent of every other lattice site. In the absence of an external electric field the probability that thermal motion will cause an ion to jump across a barrier, or a vacancy to move, is

$$p^* = A \exp(-E_b/k_B T) \tag{12-151}$$

where the height of the barrier (activation energy) is E_b and A is primarily a frequency factor (see Equation 12-157). When a weak external field is applied, the whole probability that a vacancy will jump from one site to the next in a direction parallel to the field is

$$p_t^* = p^*(\overline{E}ea/k_B T) \tag{12-152}$$

where a is the distance between lattice sites and $\overline{E}ea \ll k_B T$. The electric polarization for one jump is ea. This results in a current density, using Equation 12-152, of

$$j = np_t^*ea = np^*(\overline{E}e^2a^2/k_B T) \tag{12-153}$$

in which n is the number of vacancies per unit volume. Using Ohm's law and Equation 12-153, the conductivity is

$$\sigma = j/\overline{E} = np^*(e^2a^2/k_B T) \tag{12-154}$$

The ionic mobility is obtained from Equations 11-1 and 12-154:

$$\sigma = ne\mu = np^*(e^2a^2/k_BT) \qquad (12\text{-}155)$$

so that the mobility is

$$\mu = p^*(ea^2/k_BT) \qquad (12\text{-}156)$$

The coefficient A of Equation 12-151 which may be obtained from Equation 4-101 is considered to be a variable quantity rather than a constant (see Section 4.1.4, Volume I). This is treated in a way similar to that used in the low-temperature approximation of the Debye equation for heat capacity (Equation 4-111, Volume I).

$$A \simeq \nu(T/\theta_D)^3 \simeq (k_BT)^3/h^3\nu^2 \qquad (12\text{-}157)$$

Equation 12-157 gives the frequency factor for vacancy migration in a lattice plane parallel to the applied field. The frequency of the surrounding ions in planes parallel to $\bar{E}$ is given by ν. The denominator contains the factor ν^2 since only the longitudinal and one transverse mode of oscillations contribute to the conductivity. (The transverse mode perpendicular to the field does not affect the ionic conductivity.)

Conduction may be considered to take place by the mobility of interstitial ions or by that of the vacancies. In the case of interstitial migration Equation 12-151 gives the probability that an interstitial ion will be thermally activated to jump to another interstitial site. Where vacancies are responsible for the conduction, Equation 12-151 gives the probability that an ion will occupy an adjacent vacancy and thus effectively cause the vacancy to move. When the energy required to create a vacancy, E_1, is greater than the energy needed to form an interstitial

$$E_1 = E_I - E_L \qquad (12\text{-}158)$$

where E_I is the energy of the ion on an interstitial site and E_L is the energy of the ion on a lattice site. In this case it is probable that conduction will take place by the migration of interstitials.

The use of Equation 12-157 is bounded by two conditions. The first of these is that relatively large numbers of interstitials and vacancies exist in the lattice. For interstitial migration the energy factor of the exponent of Equation 12-151 is given by $E_T = (E_1 + E_b)$; this is the total activation energy for interstitial migration. The second condition is based upon a perfect lattice and the formation of an interstitial and a vacancy within it (the Frenkel mode). Here the total activation energy is $(E_T + E_b)/2$, since two species are created.

Where the conduction is primarily a result of vacancy migration, as is the cause for Schottky defects, the energy required for the creation of a Schottky pair, E_2, must be smaller than that for the creation of an interstitial ion. The leads to the probability that conduction in ionic crystals is primarily by vacancies rather than by interstitials if the lattice is closely packed. In this case, the activation energy for a vacancy is $(E_T + E_2/2)$, since Schottky vacancies must be made in pairs to maintain electrical neutrality.

The intrinsic conductivities for the cases just described may be summarized as follows:

1. For interstitial migration:

$$\sigma = C\exp[-(E_1 + E_b)/k_BT] \qquad (12\text{-}159)$$

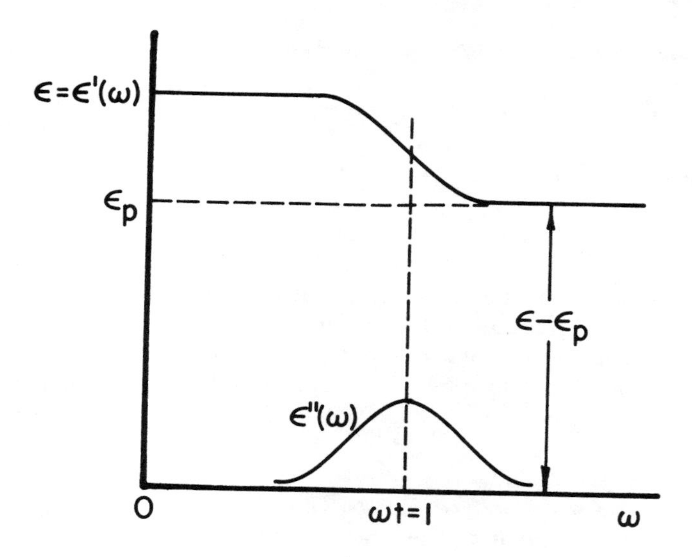

FIGURE 12-8. Schematic representations of the real and imaginary parts of the dielectric constant of a polar material, with a given relaxation time, in the microwave range.

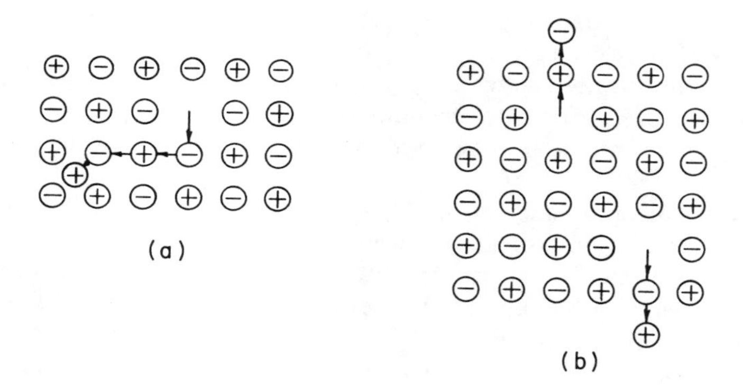

FIGURE 12-9. Point defects in ionic solids: (a) Frenkel and (b) Schottky defects.

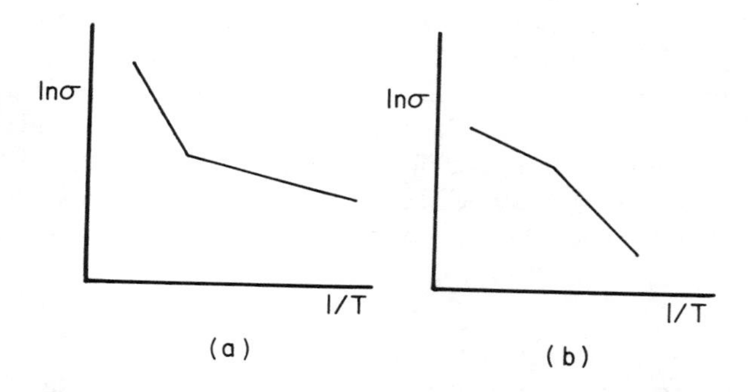

FIGURE 12-10. Schematic diagrams of the effects of various factors upon the ionic conductivity.

2. For the migration of both interstitials and vacancies:

$$\sigma = C \exp[-(E_T + E_b)/2k_BT] \quad (12\text{-}160)$$

3. For vacancy migration:

$$\sigma = C \exp[-E_T + E_2/2)/k_BT] \quad (12\text{-}161)$$

The coefficient C in each of the above equations is given by the product of the coefficient of p* as given by Equation 12-157 and the other terms in Equation 12-155: $C = n(k_BTea/\nu)^2/h^3$. The magnitude of C is about 10^6 Ω−cm)$^{-1}$.

Where conduction takes place by more than one of the means just described, the intrinsic conductivity is given by

$$\sigma = \Sigma C_i \exp(-E_i/k_BT) \quad (12\text{-}162)$$

in which the coefficients and activation energies are those appropriate for each migrating carrier.

In the case of closely packed ionic crystals, it would be expected that intrinsic conductivity should be primarily a result of vacancy migration. This should be so since, as noted previously, the concentration of Frenkel defects is very much less than that of Schottky defects. In less densely packed crystals (Section 10.6.4) interstitial migration of the smaller ion, usually the cation, becomes increasingly likely as the difference between the radii of the ions increases.

At lower temperatures the conductivity largely results from the migration of impurity ions in the lattice. In this case the summation given by Equation 12-162 would include any conduction modes by the host lattice and additional terms to account for the extrinsic behavior.

Impurities in covalent crystals frequently can account for their conductivities (Sections 11.2 and 11.2.1). Some of these impurities also may be associated with lattice defects. When the number of lattice defects is significantly greater than the number of impurity ions, a situation analogous to that described by Equation 12-160, for large numbers of interstitials and vacancies, may be employed to describe the conductivity.

Where interstitials are the majority carriers, at low temperatures, their number is small. Consequently, the number of vacancies is small. As the temperature increases, impurity ions could migrate interstitially and increase the conductivity (Figure 12-10a). If the majority carriers are vacancies at low temperatures, impurity ions could reduce the conductivity by filling the vacancies as the temperature increases (Figure 12-10b).

Other factors, in addition to those just noted, may influence the discontinuities shown in Figure 12-10. One important effect is that the activation energy decreases with increasing temperature. Another consideration is that the conduction mechanism may change. For example, if the positive ions are the majority carriers at lower temperatures and the anions become mobile at higher temperatures, the increased number of carriers could account for the behavior as sketched in Figure 12-10a. It also has been considered that vacancies may be predominant in one range of temperatures and interstitials in another. This could account for the change in conductivitity as shown in Figure 12-10b.

12.9.1.2. Lattice Treatment of Intrinsic Conductivity

Another approach to the analysis of ionic conductivity is based upon crystal structure. Starting with a one-dimensional lattice, the probability of a jump by a vacancy is

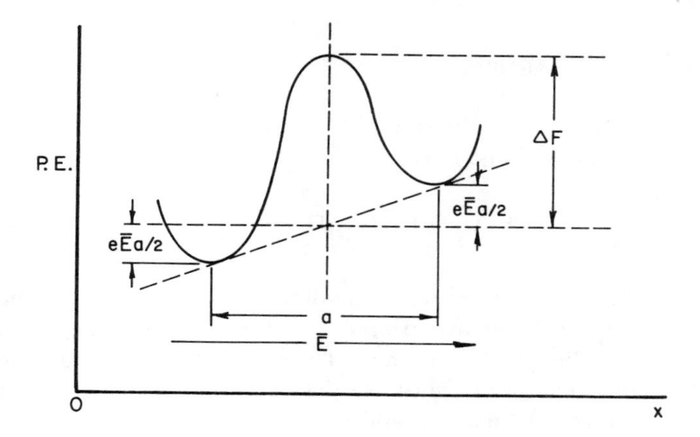

FIGURE 12-11. Change in the average potential energy of a vacancy when a uniform electric field is applied parallel to a line of ions.

$$p^* = \nu \exp(-\Delta F/k_B T) \qquad (12\text{-}163)$$

where ν is the frequency of the ions on either side of the vacancy and ΔF is the Gibbs free energy of activation in the absence of an electric field. The vacancy created by the absence of a positive ion upon a lattice site has an effective charge of −e. Upon the application of a suitable, uniform, electric field, $\bar{E}$, the vacancy may be considered to jump from one lattice site to the next (see Section 12.9.1). The application of the field increases the potential energy of the vacancy by an amount $\bar{E}ea$, where a is the interionic distance (Figure 12-11).

The probability that the vacancy will make a jump along the same direction as the field is

$$p_+^* = \nu \exp[-(\Delta F + e\bar{E}a/2)/k_B T] \qquad (12\text{-}164a)$$

since a/2 is the distance it must travel to reach the peak of the adjacent activation hump. The probability for the vacancy to jump in the direction opposite to the field is greater, being

$$p_-^* = \nu \exp[-(\Delta F - e\bar{E}a/2)/k_B T] \qquad (12\text{-}164b)$$

The average drift velocity, $\bar{V}$, of the vacancies is determined by the difference in jump probabilities, Equations 12-164b and 12-164a, and the jump distance:

$$\bar{V} = a(p_-^* - p_+^*) = a\nu \exp(-\Delta F/k_B T) \{+\exp[+(e\bar{E}a/2)/k_B T] - \exp[-(e\bar{E}a/2)k_B T]\}$$

This is more conveniently reexpressed as

$$\bar{V} = a\nu \exp(-\Delta F/k_B T)\, 2 \sinh(e\bar{E}a/2k_B T) \qquad (12\text{-}165)$$

At low external fields the hyperbolic function may be approximated as

$$2 \sinh(e\bar{E}a/2k_B T) \simeq 2\, e\bar{E}a/2k_B T = e\bar{E}a/k_B T \qquad (12\text{-}166)$$

The substitution of Equation 12-166 into Equation 12-165 gives

$$\overline{V} = (a^2\nu e\overline{E}/k_BT)\exp(-\Delta F/k_BT) \qquad (12\text{-}167)$$

for low applied electric fields. The mobility is obtained from this by means of the definition (Section 5.6.3, Volume I) as

$$\mu = \frac{\overline{V}}{\overline{E}} = (a^2\nu e/k_BT)\exp(-\Delta F/k_BT) \qquad (12\text{-}168)$$

Equations 12-167 and 12-168 are for a one-dimensional lattice. They may be applied to an NaCl-type lattice by referring to Figure 10-13b. Consider the central lattice position in this figure to be an Na^+ vacancy. This vacancy may jump to any of the 12 surrounding Na^+ sites. The presence of the field does not affect the activation energy of a jump transverse to the field. In addition, such jumps do not contribute to the flow of current. Only four of the eight remaining possible jumps which the Na^+ vacancy could make could contribute to the flow of current. Thus, for NaCl lattices, Equations 12-165 through 12-168 must be multiplied by four to convert the results for a linear lattice to those for a three-dimensional NaCl-type lattice.

The intrinsic conductivity of a material with an NaCl-type lattice can be described by an expression analogous to Equation 11-3, where the vacancies are the majority carriers, as

$$\sigma_V = n_+e\mu_+ + n_-e\mu_- \qquad (12\text{-}169)$$

in which the subscripts + and − refer to anion and cation vacancies, respectively, and n is the number of a given type of such vacancies per unit volume. Equation 12-168 is multiplied by four to adapt it to the NaCl-type lattice:

$$\mu = (4a^2\nu e/k_BT)\exp(-\Delta F/k_BT) \qquad (12\text{-}170)$$

Using Equations 12-169 and 12-170, the intrinsic conductivity is given by

$$\sigma_V = 4n_+e(a^2\nu_+e/k_BT)\exp(-\Delta F_+/k_BT) + 4n_-e(a^2\nu_-e/k_BT)\exp(-\Delta F_-/k_BT)$$

or

$$\sigma_V = 4a^2e^2/k_BT[n_+\nu_+\exp(-\Delta F_+/k_BT) + n_-\nu_-\exp(-\Delta F_-/k_BT)] \qquad (12\text{-}171)$$

If N is the total number of all possible vacancy sites of either kind for a unit volume of a crystal, the fraction of each type of vacancy, cation or anion, is given by the Boltzmann equation as

$$n/N = \exp(-F_P/2k_BT) \qquad (12\text{-}172)$$

in which F_p is the Gibbs free energy required to create a Schottky pair. It will be recalled that n_+ must equal n_- in order to preserve electrical neutrality. Multiplying the numerator and denominator of Equation 12-171 by N/N gives

$$\sigma_V = 4Na^2e^2/k_BT[(n_+/N)\nu_+\exp(-\Delta F_+/k_BT) + (n_-/N)\nu_-\exp(-\Delta F_-/k_BT) \qquad (12\text{-}173)$$

Equation 12-172 is substituted into Equation 12-173 and is factored to the form

$$\sigma_V = (4Na^2e^2/k_BT)\exp(-F_P/2k_BT)[\nu_+\exp(-\Delta F_+/k_BT) + \nu_-\exp(-\Delta F_-/k_BT)] \quad (12\text{-}174)$$

Equation 12-174 includes the contributions of both types of vacancies to the intrinsic conductivity. However, the mobility of the cation vacancies would, in general, be expected to be greater than that of the anions. This results from the relatively small sizes of the cations as compared to the anions (see Section 10.6.4). When $\mu_- \gg \mu_+$, then according to Equation 12-170, ΔF_- is much less than ΔF_+. Thus, $\exp(-\Delta F_+/k_BT)$ in Equation 12-174 is small and may be neglected for this case. For this condition Equation 12-174 becomes

$$\sigma_V \simeq (4Na^2e^2\nu_-/k_BT)\exp(-F_P/2k_BT)\exp(-\Delta F_-/k_BT) \quad (12\text{-}175)$$

and is specifically for the case in which the cation vacancies constitute the primary means of conduction.

Equation 12-175 may be reexpressed in terms of enthalpy, H, and entropy, S, by using the Gibbs equations for F and ΔF. This substitution gives

$$\sigma_V = (4Na^2e^2\nu_-/k_BT)\exp(-H_P/2k_BT + S_P/2k_B)\exp(-\Delta H_-/k_BT + \Delta S_-/k_B)$$

Like terms are combined as

$$\sigma_V = (4Na^2e^2\nu_-/k_BT)\exp[(S_P/2 + \Delta S_-)/k_B]\exp[(-H_P/2 - \Delta H_-)/k_BT]$$

or,

$$\sigma_V = (4Na^2e^2\nu_-)/k_BT)\exp[T(S_P/2 + \Delta S_-)/k_BT]\exp[-(H_P/2 + \Delta H_-)/k_BT] \quad (12\text{-}176)$$

The same restrictions hold for Equation 12-176 as for Equation 12-175. Equation 12-174, or its equivalent, must be used when both types of vacancies have significant mobilities.

12.9.1.3. Extrinsic Conductivity

The influence of impurity ions upon the electrical conductivity of ionic crystals is important. Assume that divalent impurity cations are present in an NaCl-type lattice. Each such impurity ion will bond with two anions. This results in the formation of a vacant site which normally would have been occupied by a monovalent cation. Electrical neutrality is maintained by the resultant defect structure (see Sections 10.6.7, 10.6.9, 11.7, and 11.8.4) Thus, the number of cation vacancies per unit volume is

$$n_- = n_+ + n_d \quad (12\text{-}177)$$

Here n_d is the number of vacancies per unit volume created by the presence of the divalent impurities. Since each divalent ion induces a vacant cation site, n_d equals the concentration of the divalent impurities.

It is convenient to reexpress the density of vacant anion sites as

$$n_+ = n_- - n_d \quad (12\text{-}178)$$

Using Boltzmann's equation in the same way as for Equation 12-172

$$n_+/N = (n_- - n_d)/N = \exp(-F_P/2k_BT);\; n_-/N = \exp(-F_P/2k_BT) \quad (12\text{-}179)$$

An expression involving both types of vacancies may be obtained by the product of both portions of Equation 12-179:

$$n_-/N \cdot (n_- - n_d)/N = \exp(-F_P/k_BT) \quad (12\text{-}180)$$

This may be rewritten in quadratic form as

$$n_-^2 - n_-n_d = N^2 \exp(-F_P/k_BT) \quad (12\text{-}181)$$

The number of vacancies induced by the divalent impurities is obtained from Equation 12-181 as

$$n_d = [n_-^2 - N^2 \exp(-F_P/k_BT)]/n_- \quad (12\text{-}182)$$

The solution of Equation 12-181 for n_- is useful for the description of the extrinsic conductivity of this class of solids. The use of the equation for the general solution of a quadratic equation as applied to Equation 12-181 gives

$$n_- = \frac{n_d + [n_d^2 + 4N^2 \exp(-F_P/k_BT)]^{1/2}}{2} \quad (12\text{-}183)$$

This may be reexpressed by multiplying and dividing the quantity within the brackets by n_d^2. This results in

$$n_- = \frac{n_d + \left[\dfrac{n_d^4 + n_d^2 4N^2 \exp(-F_P/k_BT)}{n_d^2}\right]^{1/2}}{2}$$

When n_d^2 is factored out of the brackets,

$$n_- = \frac{n_d}{2} + \frac{n_d}{2}\left[1 + \frac{4N^2 \exp(-F_P/k_BT)}{n_d^2}\right]^{1/2}$$

Or,

$$n_- = \frac{n_d}{2}\left\{1 + \left[1 + \frac{4N^2 \exp(-F_P/k_BT)}{n_d^2}\right]^{1/2}\right\} \quad (12\text{-}184)$$

This relationship will be used to simplify the approximation for the extrinsic conductivity.

The extrinsic conductivity may be expressed, for this case, as

$$\sigma_d = n_+e\mu_+ + n_-e\mu_- \quad (12\text{-}185)$$

Equation 12-184 now is used in Equation 12-185 to obtain

$$\sigma_d = n_+ e\mu_+ + \frac{n_d}{2}\left\{1 + \left[1 + \frac{4N^2 \exp(-F_P/k_BT)}{n_d^2}\right]^{1/2}\right\} e\mu_- \qquad (12\text{-}186)$$

When the concentration of the divalent ions is high enough to make a significant contribution to the conductivity, the number of vacancies induced by their presence will be much greater than the number of thermally induced Schottky pairs. For this case,

$$n_d^2 >> 4N^2 \exp(-F_P/k_BT) \qquad (12\text{-}187)$$

The use of Equation 12-187 in Equation 12-186 simplifies matters because the fraction within the brackets may be neglected. Thus,

$$\sigma_d \simeq n_+ e\mu_+ + \frac{n_d}{2}\,[1 + (1)^{1/2}]\, e\mu_-$$

and, finally,

$$\sigma_d \simeq n_+ e\mu_+ + n_d e\mu_- \qquad (12\text{-}188)$$

As noted previously (Section 12.9.1.1), cation mobility is much greater than that of anions in these crystals. Since $\mu_- \gg \mu_+$, it may be approximated that

$$\sigma_d \simeq n_d e\,\mu_- \qquad (12\text{-}189)$$

Corrections must be made for n_d since only a fraction contributes to the conductivity (see Equations 12-168 and 12-172).

It should be noted that situations analogous to that described here may occur in organic dielectric materials. This may take place when such substances contain impurity ions or easily ionizable molecules. Such cases permit extrinsic conduction to take place. Conditions of these kinds can result in appreciable leakage currents and eventually lead to breakdown.

Equation 12-189 may be reexpressed, for divalent impurities in an NaCl-type lattice, by using the expression for mobility given by Equation 12-170 as

$$\sigma_d \simeq n_d e(4a^2 \nu_- e/k_BT) \exp(-\Delta F_-/k_BT)$$

or

$$\sigma_d \simeq n_d(4a^2 \nu_- e^2/k_BT) \exp(-\Delta F_-/k_BT) \qquad (12\text{-}190)$$

The total conductivity is given by the sum of the intrinsic conductivity as approximated by Equation 12-175 and the extrinsic conductivity expressed by Equation 12-190.

Certain types of ionic crystals show unusually high ionic conductivities, compared to those discussed above, primarily as a result of their complex crystal structures. Frequently these are layered crystal structures in which close-packed arrays are separated by much less densely packed planes. Alkali metal ions in the latter planes can have, on a comparative basis, an extremely high degree of mobility.

Two compounds of this type which have received much attention are based on $Na_2O \cdot 11Al_2O_3$ (β alumina) and $NaAl_{11}O_{17}$ (β'' alumina). Both of these compounds contain more NaO than is shown by the formulas. The Na ions occupy sites in the less densely packed layers (sometimes called conduction planes). The O ions occupy preferred sites in the close-packed planes. This configuration permits the Na ions to have high mobilities. This results in a surprisingly high conductivity. As an example, β'' alumina can have an electrical conductivity of about 20 times that of a 0.1 *M* solution of NaCl in water in the neighborhood of room temperature.

Mobilities which result in conductivities of this magnitude have caused materials of this type to be considered as excellent candidates for solid electrolytes in batteries. In addition, these compounds are very stable and the electron component of the conductivity is very small. Other compounds such as $RbAg_4I_5$ and Li_xTiS_2 also show high conductivities. However, these materials have larger components of electron conduction than the aluminas.

The operation of batteries which use these materials as electrolytes depends upon the diffusion of the alkali metal ion across the inert electrolyte where it forms a reversible compound with the cathode. At present the use of Li anodes and cathodes (such as V_2O_5) which form reversible compounds, to permit recharging, are being investigated. Recharging is accomplished by the application of a reverse external current. Ions such as Li, K, Rb, and Ag have been used in the aluminas to replace the Na. The Li anodes are being used with Li_xTiS_2 electrolytes and V_2O_5 cathodes.

Other ionic conductors have been based upon TiS_2. Here the Li ions form layered structures by occupying planes in the TiS_2 lattice. The result is a compound of composition Li_xTiS_2. The TiS_2 layers are covalently bonded, but the intercalated Li ions are weakly held by van der Waals bonds. The resulting material has a high conductivity because of the very mobile Li ions. This material is but one of a class of compounds whose compositions are based upon TX_2, where T is a transition element and X is an atom such as S, Se, and Te.

The conduction in the layered compounds discussed above occurs in the planes occupied by the mobile ions. Other compounds show one-dimensional conduction. For example, $LiAlSiO_4$ forms a hexagonal lattice in which the Li ions are in linear arrays perpendicular to the basal planes and, thus, is unidirectional. An example of the other extreme is given by LiN (NaCl type lattice). In this case the mobility of the Li ions is high in three directions.

Batteries using solid electrolytes instead of aqueous solutions are expected to have a much longer life and a storage capacity up to about three times that of a conventional battery of the same weight. These factors also lead to the expectation of the production of very small, efficient batteries.

12.9.1.4. Ionic Conductivity and Diffusion

Where a concentration gradient exists in a solid, the net flux of ions migrating in the direction of the concentration gradient is described by Fick's first law:

$$J = -D \frac{dC}{dx}$$

J, the flux, is the number of ions which cross a unit area perpendicular to the concentration gradient per unit time, D is the diffusion coefficient (cm^2/sec), and dC/dx is the concentration gradient in the direction of interest. The migration of ions generally is considered on the basis of comparative values of D. This important factor is given by

$$D = D_o \exp(-E/k_BT) \qquad (12\text{-}191)$$

D is a measure of the quantity of the diffusing species (gram-moles), passing across an area of 1 cm^2 in 1 sec across a concentration gradient of 1 mol/cm. D_o is a frequency factor similar to Equation 12-157 and E is the activation energy for an ionic jump. It will be noted that Equation 12-191 is the same as Equation 12-151.

The ions will migrate under the influence of an external electric field as well as by thermally induced motions. Einstein's equation is based upon relating these two mechanisms. This is derived by first considering an ion of charge e in a constant electric field $\bar{E}$. The Boltzmann equation gives the concentration of the diffusing ion, at a distance x from its original position, as being proportional to $\exp(-e\bar{E}x/k_BT)$. The ion flows induced by the applied field and by thermal activation are equal at equilibrium, or

$$\mu n\bar{E} + D\frac{dn}{dx} = 0$$

Since

$$n = A\exp(-e\bar{E}x/k_BT)$$

and

$$dn/dx = -(Ae\bar{E}/k_BT)\exp(-e\bar{E}x/k_BT)$$

then, by substitution in the equation for equilibrium conditions,

$$\mu \cdot A\exp(-e\bar{E}x/k_BT)\bar{E} = D(Ae\bar{E}/k_BT)\exp(-e\bar{E}x/k_BT)$$

The quantity $A\bar{E}\exp(-e\bar{E}x/k_BT)$ vanishes and the equation reduces to

$$\mu = De/k_BT$$

or,

$$\frac{\mu}{D} = \frac{e}{k_BT} \qquad (12\text{-}192)$$

This is the Einstein equation of interest for the case under consideration. Both sides of Equation 12-192 are multiplied by ne, n being the number of carriers per unit volume, to give

$$\frac{ne\mu}{D} = \frac{ne^2}{k_BT} = \frac{\sigma}{D} \qquad (12\text{-}193)$$

This conveniently relates the conductivity to the diffusion coefficient.

Equation 12-193 does not always agree with experimental data because some jumps do not contribute to the current flow. This may be corrected in a manner similar to the way in which Equation 12-170 was obtained from Equation 12-168 (see Section 12.9.1.2). The agreement, however, improves as the temperature increases. At relatively low temperatures the normally low, intrinsic mobility is enhanced by the presence of crystalline defects. At higher temperatures, the mobility is much greater and is considerably less affected by imperfections. This behavior is similar to that inducated in Figure 12-10a, the portion of the curve with the larger slope corresponding to Equation 12-193. This corresponds to the range of temperatures at which ionic diffusion appreciably increases because D increases.

In most dielectric materials the direct current decays with time. That is, when these crystals are subjected to constant voltages, at intermediate temperatures, the d.c. current diminishes to a lower, steady-state value, with a residual d.c. resistance R_r. This decrease in current, which occurs at a high initial rate, makes experimental measurement of the conductivity difficult. The initial flow of current also results in the formation of a space charge. The initial resistance is $R_i = V/I_i$, where I_i is the initial current. (The true d.c. resistance, R_t, which represents the inherent property of the material, may be determined from the sum of the conduction and capacitive currents.) See Equations 12-198 and 12-201. The initial and true resistances are rarely equal. This has been explained by the formation of the space charge which forms as a result of the flow of ion carriers in the material.

The space charge may be determined from the variation of the potential within the dielectric material. This potential, V_p, defines the field, $\overline{E}_p$, as

$$\overline{E}_p = -\frac{\partial V_p}{\partial x} \tag{12-194}$$

Considering the ion flow in the x direction, the number of diffusing carriers per unit volume is

$$n_p = \beta \exp(-eV_p/k_BT) \tag{12-195a}$$

where β is a constant. Equation 12-195a may be converted into a more useful form by means of the Einstein relationship (Equation 12-192). Both sides of Equation 12-192 are multiplied by V_p to give $V_p\mu/D = eV_p/k_BT$. This is used to reexpress Equation 12-195a as

$$n_p = \beta \exp(-V_p\mu/D) \tag{12-195b}$$

The potential, V_p, does not vary uniformly within the dielectric. Some analyses make use of Poisson's equation, which relates V_p to the charge density, rather than Equation 12-194, for this reason. In addition, the direction of $\overline{E}_p$ is opposite to that of the applied field, $\overline{E}$. This causes the migration of some of the carriers in a direction opposite to that caused by $\overline{E}$ as determined by Fick's law. This is known as back diffusion. The current density is the resultant of these two mechanisms:

$$j = \sigma\overline{E} - De\frac{\partial n_p}{\partial x} = ne\mu\overline{E} - De\frac{\partial n_p}{\partial x} \tag{12-196}$$

Equation 12-195b is a solution to Equation 12-196 when equal numbers of ion carriers flow in opposite directions. The variation in j, at a given temperature, depends upon the variation of $\partial n_p/\partial x$ in the space-charge region. Experiments show that smaller space charges are induced in purer materials.

When the space charge fully develops, the d.c. resistance remains constant for a short time. Then, as diffusion progresses, the carriers are swept out of the dielectric, the resistance approaches a very high value, and the current decreases. These mechanisms are shown schematically in Figure 12-12.

In terms of Equation 12-196, the initial current density is relatively high prior to the formation of space charge. The formation of the space charge requires times of the order of 5 msec. Once this is formed, back diffusion takes place and the current density is decreased. The relatively constant current density results from the transport of impurity ions and may exist for times ranging from seconds to minutes, depending upon

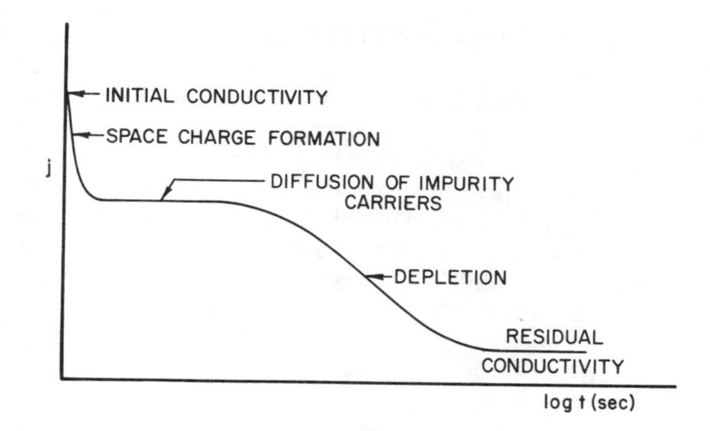

FIGURE 12-12. Schematic graph of a d.c. current density as a function of time in a dielectric. (j is of the order of 10^{-12}amp/cm^2.)

the degree of purity. The removal of the impurity is virtually completed in times of the order of a few minutes. After this interval, the conductivity and thus the current density, drops to a residual level.

The field of the space charge is opposite to that of the external field. This imparts an internal potential, V_p, opposite to that of the applied potential, V. This effect may be taken into account using Ohm's law for the initial current

$$I_i = \frac{V}{R_i} \tag{12-197}$$

where R_i is the initial resistance. And, for the residual current.

$$I_r \simeq \frac{V - V_p}{R_t} \tag{12-198}$$

in which R_t is the true resistance. A first approximation for $V_{p(max)}$ may be obtained which is based upon very pure materials. The difference between Equations 12-197 and 12-198 is

$$I_i - I_r \simeq \frac{V}{R_i} - \frac{V - V_p}{R_t} \tag{12-199}$$

For very pure materials $R_i \simeq R_t$ so that Equation 12-199 reduces to

$$I_i - I_r \simeq \frac{V_{p(max)}}{R_i} \; ; \; V_{p(max)} \simeq (I_i - I_r)R_i \tag{12-200}$$

Substituting R_i from Equation 12-197 into Equation 12-200 gives

$$V_{p(max)} \simeq (I_i - I_r)V/I_i \tag{12-201}$$

The use of the initial and residual currents to determine $V_{p(max)}$ makes it much easier to approximate the true resistance, R_t, using Equation 12-198.

12.10. EFFECTS OF STRONG FIELDS AND BREAKDOWN

The previous discussions show that the conductivities of dielectrics are affected by their purity. Ohm's law is obeyed up to high field strengths when V_p (Equation 12-198) is taken into account. This relationship between current density and field strength no longer holds when the energy exerted by the field approximates the thermal energy required for an ion jump. Field strengths of the order of 10^5 V/cm are required for this in pure crystals. Fields of this magnitude approach those required for breakdown.

Beyond the range of applicability of Ohm's law the conductivity may be approximated by the empirical equation

$$\sigma = A\exp(B\bar{E}) \qquad (12\text{-}202)$$

in which A and B are constants at a given temperature. The exponential coefficient B decreases with increasing temperature because the bonding energies also decrease and the space charges break down. In addition, intrinsic conductivity may make a significant contribution under these conditions.

Large increases in the electrical conductivity of dielectrics would be expected when such materials are subjected to very strong fields. Almost all crystalline dielectric materials contain some impurities and some show photoelectric properties (Sections 11.2.1, 11.3 and 11.8.1). This means that it is possible for a very small number of electrons to occupy states in the conduction bands of such materials. Under the influence of very strong fields these few, nearly free electrons would be accelerated so that they could promote other electrons to the conduction band in a manner similar to the avalanche mechanism described in Section 11.6.1. This process could result in a relatively large density of electrons in the conduction band. The resulting large increase in current, caused by ionic as well as electronic conductivity, then would lead to breakdown (sometimes called conductive breakdown or dielectric strength).

A simple model for this behavior, due to Fröelich, is given here for illustrative purposes. The drift velocity of an electron in the conduction band is obtained from Equation 5-61 (Volume I) as

$$\bar{V} = \bar{E}e\tau(E)/m^* \qquad (12\text{-}203)$$

in which $\tau(E)$ is the relaxation time of an electron of energy E. The energy of such an electron is

$$E = m^*\bar{V}^2/2 = (m^*/2)\,(\bar{E}e\tau(E)/m^*)^2 \qquad (12\text{-}204)$$

The rate of change of the energy of the electron as a result of the application of the electric field is obtained by differentiating Equation 12-204 with respect to time:

$$dE/dt = \bar{E}^2 e^2 \tau(E)/m^* \qquad (12\text{-}205)$$

If that portion of the energy generated by this mechanism which is conducted out as thermal losses is greater than the energy required to initiate avalanching, a small leakage current can flow without inducing breakdown; large numbers of electrons will not be excited to the conduction band. This condition is expressed as

$$-(dE/dt)_{th} > \bar{E}^2 e^2 \tau(E)/m^* \qquad (12\text{-}206)$$

$(dE/dt)_{th} = dU/dt$ (Equation 12-150) becomes high at frequencies where $\varepsilon''(\omega)$ is large.

Therefore, when the energy balance is such that the heat is removed too slowly,

$$(dE/dt)_{th} = \bar{E}^2 e^2 \tau(E)/m^* = dU/dt \qquad (12\text{-}207)$$

breakdown will occur for all energies greater than that required to promote a bound electron to the conduction band.

Any condition which diminishes τ(E) will enable breakdown to occur at higher field strengths. It will be recalled that increased temperature has the effect of decreasing τ(E) (Section 6.2 and Equations 6-3 and 6-4, Volume II). More advanced theories indicate that electron avalanching takes place at field strengths of about 80% of that given in Equations 12-206 and 12-207. This results in breakdown at correspondingly smaller fields than are required by the above equations.

Another type of breakdown can take place in nonhomogeneous dielectric materials. These substances will contain volumes of material with lower electrical resistivities than the surrounding matrix. The low-resistivity volume will conduct more leakage current than the matrix. The resistivity is decreased further as the temperature and current increase and the localized heating increases. A temperature finally is reached at which the conduction is large enough to cause breakdown. This type of thermal breakdown may be encountered in multiphase polymeric dielectrics.

Other solid dielectric materials may break down as a result of molecular dissociation under the influence of strong electric fields. In cases where other readily ionized substances, or other impurity ions, are present in a polymeric dielectric, leakage currents may be caused in a way analogous to the extrinsic behavior described in Section 12.9.1.3. If the concentration of the impurity ions is excessive, either as a result of a high, initial impurity concentration or by the continuous dissociation of the matrix or of the ionizable impurities, leakage currents will be present and breakdown can occur.

Breakdown also may be primarily mechanical in nature. As noted in Sections 12.2, 12.3, and 12.13, the application of an external field can cause both ionic and dipolar displacements. When the field strength becomes sufficiently higher, a large compressive force can be induced in the dielectric material. A point may be reached at which this force exceeds the compressive strength of the material. This results in mechanical failure and consequent breakdown.

Voids may be present in dielectric solids as a result of evaporated solvents, moisture, or improper manufacturing techniques. These voids may act as stress concentrators. Thus, mechanical failure may occur at lower strength applied fields than in the case when such voids are absent.

12.11. OPTICAL PROPERTIES

Some of the optical behavior of dielectric materials were briefly noted in Section 12.6, in conjunction with the effects of polarizability and polarization. The optical properties are examined more directly in this section in terms of the dielectric constant. This is possible because the Maxwell equation, $\varepsilon = n^2$ (Equation 12-39) provides a simple, direct relationship between the dielectric constant and the index of refraction.

The least complicated expression for the dielectric constant may be obtained starting with Equation 12-9. This is given again, for convenience, as

$$P = \frac{\bar{E}}{4\pi} (\epsilon - 1) \qquad (12\text{-}208)$$

where P is the polarization and $\bar{E}$ is the applied field. Equation 12-208 is rearranged to give the dielectric constant

$$\epsilon = 1 + \frac{4\pi}{\bar{E}} P \qquad (12\text{-}209)$$

An expression for P is obtained from Equation 12-106 which is written more generally as

$$P = \frac{Ne^2\bar{E}}{m(\omega_o^2 - \omega^2)} \qquad \text{where } \alpha = \frac{e^2}{m(\omega_o^2 - \omega^2)} \qquad (12\text{-}210)$$

Equation 12-210 is substituted into Equation 12-209, giving

$$\epsilon = 1 + \frac{4\pi}{\bar{E}} \cdot \frac{Ne^2\bar{E}}{m(\omega_o^2 - \omega^2)} \qquad (12\text{-}211)$$

The factor $\bar{E}$ vanishes and

$$\epsilon = 1 + \frac{4\pi Ne^2}{m(\omega_o^2 - \omega^2)} \qquad (12\text{-}212)$$

Equation 12-212 also may be expressed in terms of the polarizabilty, α, as given by Equation 12-210:

$$\epsilon = 1 + 4\pi N\alpha \qquad (12\text{-}213)$$

A plot of ε as a function of ω is given in Figure 12-13a; this also shows the behavior of the index of refraction, since $\varepsilon = n^2$.

Equation 12-212 represents the case where little or no damping is present, since this factor was not taken into account in the derivation of Equation 12-106, given here as Equation 12-210. As such, it provides insight into the behavior of the index of refraction as a function of frequency for those ranges of frequency in which negligible damping occurs. In using Equation 12-212 in this way, care must be taken to use the appropriate value for α in the range being considered (Figures 12-7 and 12-13b).

The value of ε increases slowly as the frequency increases (Figure 12-13a). The index of refraction is real since ε is positive. Under these conditions dielectrics are transparent and usually have refractive indexes greater than unity. As the frequency approaches ω_o, the index of refraction increases rapidly and dispersion (change of index of refraction with wavelength) occurs in this range. A singularity exists at $\omega = \omega_o$. These can occur in the microwave, infrared, and ultraviolet ranges for many materials, depending upon the mechanisms contributing to the polarizability (see Figure 12-13b). At frequencies slightly greater than ω_o, ε is large and negative. The magnitude of $-\varepsilon$ decreases as ω increases and is zero where the fraction in Equation 12-212 equals -1. The index of refraction is imaginary in this range; total reflection of the given type of radiation takes place. At those frequencies at which ε again becomes positive, n is real and the material is transparent once more to radiation of higher frequencies. As noted in connection with the application of Equation 12-106, Equation 12-212 provides a reasonable approximation when ω does not approach the natural frequency.

However, damping must be taken into account at the resonant frequency of each of the dipolar, atomic, and electronic components. The energy absorption, or losses, in the microwave region are proportional to Equation 12-139b. That which occurs in the infrared range may be approximated in a similar way by considering only P_a and P_e. Since the relaxation time for $P_a \gg P_e$ in this range of frequencies, it may be approx-

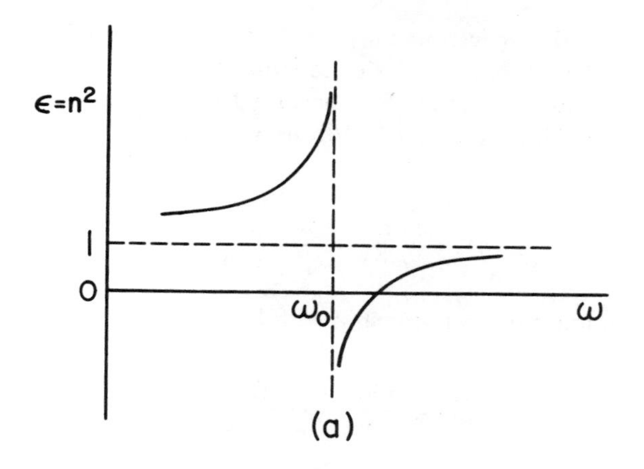

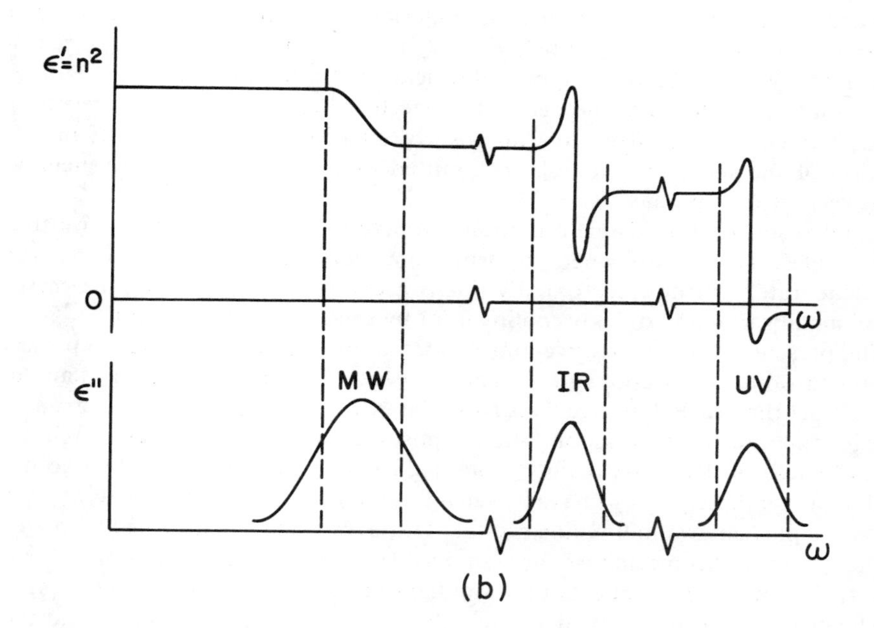

FIGURE 12-13. (a) The dielectric constant and index of refraction as functions of frequency, in the range of a resonant frequency, for the case of little or no damping and (b) where damping occurs at resonant frequencies (see Figures 12-7 and 12-8).

imated, using the same approach as for Equations 12-139a and 12-139b, that the dielectric constants in the infrared range are

$$\epsilon'(\omega) = \epsilon_\infty + \frac{\epsilon - \epsilon_\infty}{1 + \omega^2\tau^2}$$

and

$$\epsilon''(\omega) = (\epsilon - \epsilon_\infty)\,\frac{\omega\tau}{1 + \omega^2\tau^2}$$

in which ε_∞ is the dielectric constant due to electroinc polarization and ε is the dielectric constant resulting from all sources of polarization.

The visible range of the spectrum may be described by relationships based upon Equations 12-116a and 12-116b. It will be recognized that the damping coefficient, B, is very small in this frequency range. This permits the approximation that B^2 is negligible. Then, Equations 12-116a and 12-116b may be written for the range of visible frequencies as

$$\epsilon_\infty' \simeq \frac{4\pi Ne^2}{m} \frac{\omega_0^2 - \omega^2}{(\omega_0^2 - \omega^2)^2} + 1$$

(It is interesting to compare this relationship with Equation 12-34.) And

$$\epsilon_\infty'' \simeq \frac{4\pi Ne^2}{m} \frac{2B\omega}{(\omega_0^2 - \omega^2)^2}$$

The variation of the dielectric constant as a function of frequency is shown schematically in Figure 12-13b. Dielectric materials are transparent in those portions of the spectrum where ε'' is negligible; virtually no absorption takes place since resonance occurs only where ε'' becomes appreciable near a natural frequency.

It should be noted that the nearly free electrons in bulk metals are responsible for their opaqueness. The electrons can absorb radiation at all frequencies in the visible portion of the spectrum; none is transmitted. Very thin (sub-micron) metallic films, however, are transparent.

Some absorption of visible radiation can take place in the normally transparent alkali halides. This occurs when impurities and imperfections are present naturally or are deliberately introduced. Usually this is accomplished by heating the crystal in an alkali metal vapor and quickly cooling it, or by exposure to X-radiation.

The presence of extra positive ions creates anion vacancies in order to preserve electrical neutrality (see Sections 12.9.1 and 12.9.1.2). The positively charged anion vacancies attract the nearly free electrons from the extra metal ions. These electrons are very mobile and can resonate among the six positive ions which surround each anion vacancy. These electrons are readily promoted to the conduction band by photons in the visible range; these photons have wavelengths lying between 4 and 6×10^3 Å, depending upon the particular alkali halide crystal. Those wavelengths absorbed by the electron-vacancy pair in this mechanism are removed from the incident radiation. This selective absorption of the components of white light gives rise to the colors of these crystals.

The electron-vacancy pair described above is known as an F center and the corresponding radiation absorption bands are called F bands. Color centers also may be induced by electron, γ-ray, and neutron irradiation.

Combinations of adjacent F centers also affect the optical properties of these crystals. Two and three such F centers, known respectively as M and R centers, absorb visible radiation in a more complex way than does a single F center. Another type of color center is the F_A center. It consists of an anion vacancy in a NaCl-type lattice surrounded by five of the normally present alkali metal ions. The sixth nearest neighbor is an alkali metal ion of another species. F_A centers have two absorption bands.

Trapped holes also may form color centers. A halogen ion with a trapped hole lacks one electron in its outer shell. This ion interacts with an adjacent halogen ion to form a halogen molecule which lacks one electron; this combination can be considered to be a singly ionized halogen molecule. Such a center is called a V_K center.

12.12 FERROELECTRIC CRYSTALS

Up to this point materials have been considered in which the alignment of randomly oriented dipoles in a substance results only from the application of an applied electric

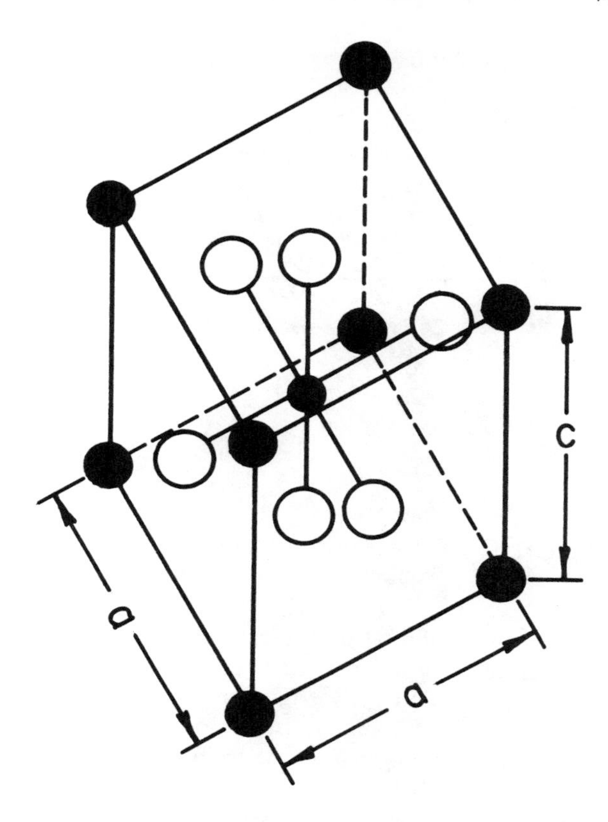

FIGURE 12-14. Tetragonal $BaTiO_3$ structure cell: Ba^{+2} ions on corner sites, O^{-2} ions on face sites and a Ti^{+4} ion slightly offset from the central site; $c/a \simeq 1.01$.

field. Such polarization is a linear function of $\bar{E}$ (Equation 12-9). The properties of ferroelectric crystals result from a spontaneous polarization of their components and domains are formed in the absence of an external field. Their polarizations also may be made to show hysteresis loops by the application of electric fields. In addition, their spontaneous polarizations vanish at a critical temperature at which they become paraelectric. Thus, a strong resemblance exists between the nature and properties of ferroelectric and ferromagnetic materials.

These properties usually result from a crystal structure in which a cation is slightly displaced from the center of negative charge in a structure cell. In other words, their lattices are not exactly symmetrical.

Three general classes of ferroelectrics have been observed. The simplest type, the perovskites, of which $BaTiO_3$ is best known, is one in which the general formula is MTO_3. M is a normal metal ion, T is the transition element, and O is an oxygen ion. $PbTiO_3$ and $KNbO_3$ are other examples of this class. Here, the transition ion occupies a position slightly offset from the center of a tetragonal lattice. The O ions in octahedral positions furnish the negative charge (Figure 12-14).

A second class, typified by potassium dihydrogen phosphate (KH_2PO_4, also frequently designated as KDP) is one in which hydrogen bonds induce polarization. In these materials the hydrogen ions (protons) occupy certain ordered lattice sites and, depending upon their location, contribute to the polarization. The spontaneous polarizations of ferroelectrics based upon hydrogen bonds are about 10% of those of the perovskites.

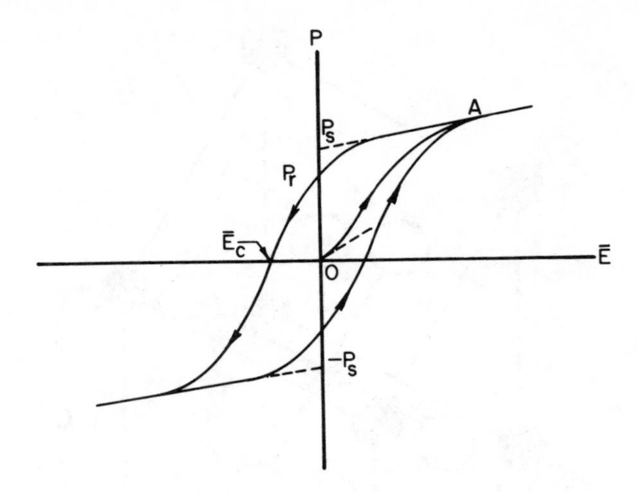

FIGURE 12-15. Polarization of a ferroelectric crystal in an external field. Note that A is the saturation polarization.

A third type consists of salts similar to Rochelle salt, $NaKC_4H_4O_6 \cdot 4H_2O$. The crystal structures of these salts are very complex and the displacements responsible for their ferroelectric properties are still under investigation.

As is the case for ferromagnetic materials, the polarization in ferroelectrics varies in direction from domain to domain within the solid. Their high degree of polarizability results in high values of ε. This is responsible for the utility of ferroelectrics and is an important factor in the miniaturization of many electrical devices.

The polarization of a ferroelectric material as a function of an applied field is shown in Figure 12-15. Starting with a ferroelectric crystal which has randomly oriented domains, where the net polarization is zero, the application of an electric field causes those domains which are oriented parallel to the field to absorb adjacent, less-favorably oriented domains and the polarization increases. The mechanism of this reaction differs considerably from that of ferromagnetic domains. In addition, under certain conditions, favorably oriented domains may grow only by nucleation and not by wall motion.

At this point, it is necessary to describe the properties of ferroelectric domains. These may be observed by the application of films of fine, charged powders (analogous to Bitter patterns), etching, and by other methods. Domains exist because the resultant reduction of the electrostatic energy of the spontaneous polarization lowers the free energy (see Section 9.6, Volume II). As in the ferromagnetic case, a limit to the decrease in domain size is imposed when the size decrease, caused by the lowering of the electrostatic energy, becomes equal to the energy required to create additional domain walls. The domain wall thicknesses are of the order of a few lattice parameters. These are much smaller than Bloch walls which are of the order of 100 times larger (Equation 9-54, Volume II).

The small domain wall width results from the minimization of the stored electrostatic and anisotropy energies. The anisotropy energy is important here because such energy differences between oppositely oriented domains is very small. No factor corresponding to the exchange energy (normally about ten times the anisotropy energy) is present. This causes the interaction energies for oppositely oriented dipoles to be small and to be very close in magnitude. This situation, in addition to the absence of an analogue of the exchange energy, explains the thin walls of ferroelectric domains.

Another interesting aspect of ferroelectric domain walls is their difference from

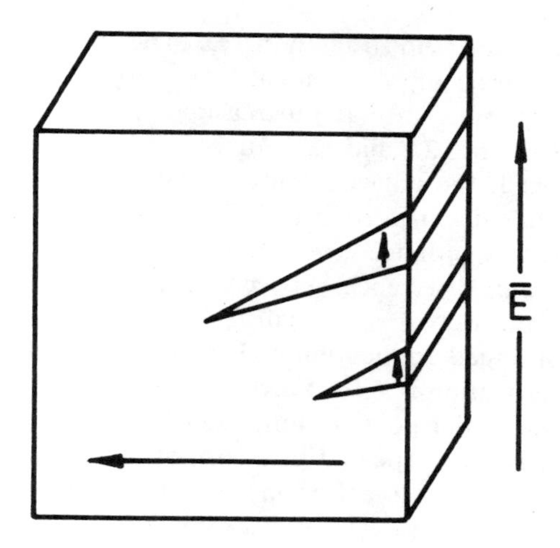

FIGURE 12-16. Nucleation of new domains in a ferroelectric crystal. The arrows indicate the polarizations of the host and nucleating domains.

Bloch walls (Section 9.7, Volume II). Where the anisotropy of a ferroelectric crystal containing a domain wall is high (structure cell with low symmetry), the polarization vector decreases in magnitude without undergoing rotation across the domain wall, goes through zero, and then increases in magnitude in the direction of the vector of the adjacent domain. The anisotropy energy determines the structure of the domain wall, but does not add to its energy. In crystals with very high symmetry (very low anisotropy energies) the polarization vectors can rotate across the domain wall just as the spin vectors do in the Bloch model.

At very low fields (of the order of 200 to 300 V/cm) the increase in polarization (Figure 12-15) as a function of time can be a result of the nucleation and initial growth of incompletely formed domains. These always nucleate at the surface (Figure 12-16). The wedge-shaped nuclei grow in both the longitudinal (forward) directions as well as in the transverse (sidewise) directions.

The nucleation rate for domain formation is given by

$$dn/dt = A\exp(-\bar{E}_{cr}/\bar{E}) \qquad (12\text{-}214)$$

in which A is a constant and $\bar{E}_{cr}$ is the critical electric field strength. It also has been observed that the velocity of domain wall motion in low electric fields is expressed by

$$v = v_{\infty}\exp(-\bar{E}'_{cr}/\bar{E}) \qquad (12\text{-}215)$$

over a velocity range from approximately 0.01 to 0.1 cm/sec. The constant v_{∞} is about 10 cm/sec for $BaTiO_3$. The similarity between Equations 12-214 and 12-215 suggests that the polarization may increase at low fields by domain nucleation and by wall motion.

The initial polarization (Figure 12-15) is approximately linear. Virtually no domain growth or nucleation take place. As the applied field increases, domain growth occurs by wall motion and nuclei formation. These mechanisms operate, with increasing external field strength, until the entire crystal becomes a single domain at point A, where saturation is reached. As the field strength increases beyond this point, the polarization increases as a result of larger dipole moments. The larger dipole moments result from increased displacements induced by the increasingly greater applied field (see Section 8.6, Volume II, and Sections 12.2 and 12.5.3).

As the field is reduced, the polarization decreases linearly for a given range of $\bar{E}$. The extrapolation of this linear portion to $\bar{E} = 0$ is the spontaneous polarization, P_s. The polarization decreases with decreasing $\bar{E}$ until at $\bar{E} = 0$ the remanent polarization, P_r, is reached. The small difference between P_s and P_r indicates that the crystal is virtually a single domain; only a very small fraction of the volume has reverted to its original state. The continued application of $\bar{E}$ reduces the polarization until at $\bar{E}_c$, the coercive field, the polarization is zero. Further application of $\bar{E}$ results in polarization opposite to that of the initial polarization; the direction of the dipole moments has reversed. The process is then repeated in this opposite direction in a way paralleling that of ferromagnetic hysteresis (see Section 9.8 and Figure 9-15a, Volume II).

12.12.1. Theory of Ferroelectrics

The local field responsible for ferroelectric properties is given by a more general form of Equation 12-25 as

$$\bar{E}_{loc} = \bar{E} + \beta P \tag{12-216}$$

where β is the Lorentz internal field constant; $\beta = 4\pi/3$ for a spherical cavity. For temperatures above the Curie temperature P is the total polarization as given by $P_a + P_e$. This was derived from the Langevin function (Section 8.3.1, Volume II) and from Equation 12-20 is

$$P = \frac{Np_d^2 \bar{E}_{loc}}{3k_BT} \tag{12-20}$$

The substitution of Equation 12-216 into Equation 12-20 gives

$$P = \frac{Np_d^2}{3k_BT}(\bar{E} + \beta P)$$

This may be rewritten as

$$P = \frac{Np_d^2 \bar{E}}{3k_BT} + \frac{\beta P N p_d^2}{3k_BT}$$

Upon rearrangement

$$P - \frac{\beta P N p_d^2}{3k_BT} = P(1 - \beta N p_d^2/3k_BT) = \frac{Np_d^2 \bar{E}}{3k_BT}$$

Then the polarization is

$$P = \frac{\dfrac{Np_d^2 \bar{E}}{3k_BT}}{1 - \dfrac{\beta N p_d^2}{3k_BT}} \tag{12-217}$$

And the ferroelectric susceptibility in the paraelectric range ($T > T_c$) is

$$\chi_f = \frac{P}{\bar{E}} = \frac{\dfrac{Np_d^2}{3k_BT}}{1 - \dfrac{\beta Np_d^2}{3k_BT}} \quad (12\text{-}218)$$

Equation 12-218 may be expressed in the Curie-Weiss form, as in Equation 9-8 (Volume II), by letting $T_c = \beta Np_d^2/3k_B$ (see Equation 12-229). This results in

$$\chi_f = \frac{T_c/\beta}{T - T_c} \quad (12\text{-}219)$$

And, by use of Equation 12-10,

$$\epsilon = \frac{4\pi P}{\bar{E}} + 1 \simeq \frac{4\pi P}{\bar{E}} \simeq 4\pi\chi_f \simeq \frac{4\pi T_c/\beta}{T - T_c} \quad (12\text{-}219a)$$

Or, when Equation 12-229 is taken into account

$$\chi_f = \frac{Np_d^2/3k_B}{T - T_c} \quad (12\text{-}219b)$$

And, similarly

$$\epsilon \simeq \frac{4\pi Np_d^2/3k_B}{T - T_c} \quad (12\text{-}219c)$$

Insight into spontaneous polarization for $T < T_c$ may be obtained by the use of an expression based upon Equation 8-50 (Volume II). Starting with

$$P = Np_d < \cos\theta > = Np_d L(a); \; a = p_d \bar{E}_{loc}/k_BT \quad (12\text{-}220)$$

The inclusion of Equation 12-216 for $\bar{E}_{loc}$ gives

$$P = Np_d L[(p_d/k_BT)(\bar{E} + \beta)] \quad (12\text{-}221)$$

For spontaneous polarization, $\bar{E} = 0$. So, using an adaptation of Equation 8-49 (Volume II),

$$P_s = Np_d[(p_d/k_BT)\beta P] = \frac{Np_d^2\beta P}{k_BT} \quad (12\text{-}222)$$

Equation 12-222 equals the saturation polarization, P_s, which occurs in that portion of the Langevin function where L(a) becomes asymptotic. When this is the case, virtually all of the dipoles are spontaneously aligned. Under these special conditions

$$P_s = \frac{Np_d^2\beta P}{k_BT} \simeq P_S \quad (12\text{-}223)$$

Here $P \rightarrow P_s$ so that Equation 12-223 becomes

$$1 = \frac{Np_d^2\beta}{k_BT} \simeq \frac{P_S}{P_s} \qquad (12\text{-}224)$$

or

$$P_s \simeq P_S \qquad (12\text{-}224a)$$

Since this represents the maximum polarization, P_S/P_s may be approximated as being close to unity. So the maximum polarization is approximated from Equation 12-224 as

$$P_S \simeq \frac{Np_d^2\beta}{k_BT} \qquad (12\text{-}225)$$

Lesser degrees of polarization are given by

$$P_s = L(a) = \frac{p_d\bar{E}_{loc}}{k_BT} = \frac{p_d(\bar{E} + \beta P_s)}{k_BT} = \frac{p_d\beta P_s}{k_BT} = a \qquad (12\text{-}226)$$

The relative polarization is obtained from the use of Equations 12-225 and 12-226 as

$$P_{rel} = \frac{P_s}{P_S} \simeq \frac{k_BT}{Np_d^2\beta}\, a \qquad (12\text{-}227a)$$

Or, substituting for the Langevin factor, a,

$$P_{rel} \simeq \frac{k_BT}{Np_d^2\beta}\,\frac{p_d\beta P}{k_BT} = \frac{P}{Np_{d(max)}} \qquad (12\text{-}227b)$$

The values of P_{rel} for various degrees of polarization may be obtained by graphical solutions of Equation 12-227a which is a linear function of a and the Langevin function. This is shown schematically in Figure 12-17 and is analogous to Figure 9-1 (Volume II).

Equation 12-227a is tangent to L(a) in the range in which L(a) approaches zero. This is the equivalent of saying that their slopes are equal. The slope of the Langevin function, obtained from the derivative of Equation 8-49 (Volume II), is $dL(a)/da = 1/3$. The slope of Equation 12-227a is

$$\frac{dP_{rel}}{da} = \frac{k_BT}{Np_d^2\beta} \qquad (12\text{-}228)$$

Equating the slopes of Equation 12-227a and L(a):

$$\frac{k_BT}{Np_d^2\beta} = \frac{1}{3}$$

and the Curie temperature is found to be

$$T_c = \frac{Np_d^2\beta}{3k_B} \qquad (12\text{-}229)$$

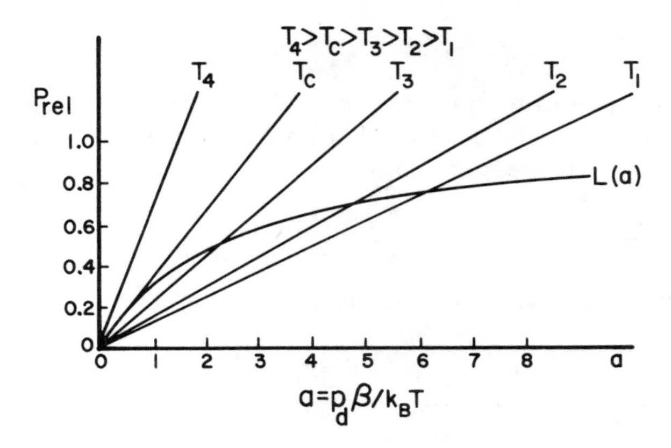

FIGURE 12-17. Relative amounts of spontaneous ferroelectric polarization as functions of temperature.

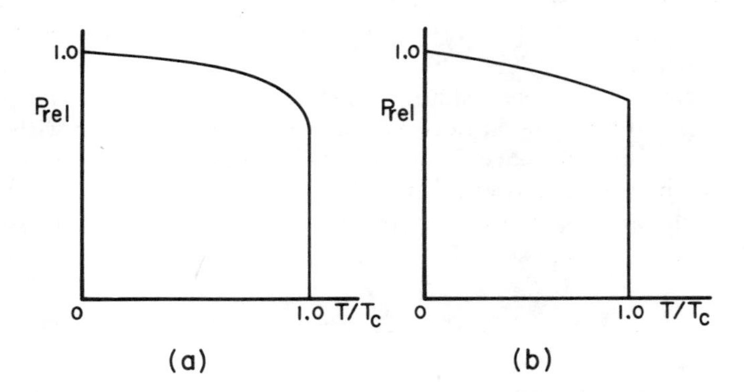

FIGURE 12-18. (a) General behavior of the relative polarization normalized with respect to the Curie temperature. (b) Discontinuous behavior.

T_c is that temperature at which spontaneous polarization vanishes and the material becomes paraelectric (see Section 9.1, Volume II). The expression for the Curie temperature obtained in Equation 12-229 is the same as that assumed from Equation 12-218 and applied in Equation 12-219. The behavior of the relative ferroelectric polarization as a function of temperature, normalized with respect to T_c, is shown in general ways in Figure 12-18.

An undesirable consequence of the use of the Lorentz field in Equation 12-220 is that while it simplifies the analysis, the values obtained for P_S (Equation 12-225) and T_c (Equation 12-229) may be higher than the experimental values for some ferroelectrics. This difficulty may be overcome by the use of the Onsager cavity field (Equation 12-55). However, this should be used with caution, especially where ε is large.

The dielectric constants of ferroelectric crystals above the Curie temperature behave similarly to those discussed for crystals which do not show spontaneous dipole alignment. This may be shown starting with Equation 12-76.

$$\epsilon = \frac{1 + 2P}{1 - P} \tag{12-76}$$

where P represents the combined polarization effects of ionic and electronic polarizations. Now applying Equation 12-17 in a more general form

$$P = Np = N\alpha\bar{E}_{loc} \tag{12-230}$$

Here, because of the constitution of P, α must also represent the combined effects of ionic and electronic polarizability. The Clausius-Mossotti equation Equation 12-38 is convenient to apply here in the form

$$P = \frac{4\pi}{3} N\alpha \tag{12-231}$$

Equation 12-231 now is substituted into Equation 12-76 and results in

$$\epsilon = \frac{1 + 2P}{1 - P} = \frac{1 + \frac{8\pi}{3} N\alpha}{1 - \frac{4\pi}{3} N\alpha} \tag{12-232}$$

The condition for which $N\alpha = 3/4\pi$ is the critical value responsible for what is also known as the polarization catastrophe. This case corresponds to a very large, but finite, polarization in the absence of an applied field. The tendency for $N\alpha \rightarrow 3/4\pi$ in ferroelectric materials accounts for their high values of ε and, consequently, their utility in electrical devices.

The dielectric constants of perovskites, in the paraelectric state, can be expressed in terms of changes in their cubic lattices as functions of temperature without recourse to the assumption of the presence of permanent dipoles. The derivation given below is based upon that of Jonker and van Santen.[16]

This relationship may be shown by starting with the Clausius-Mossotti equation (Equation 12-38) as applied to structure cells:

$$\frac{\epsilon - 1}{\epsilon + 2} = \frac{4\pi}{3} N\alpha = AN; \quad A = \frac{4\pi}{3} \alpha \tag{12-233}$$

where α is considered to be independent of temperature and equals $\alpha_i + \alpha_e + \alpha_{corr}$; α_{corr} is a correction term. N is the number of structure cells per cm^3. The differentiation of Equation 12-233 with respect to temperature results in

$$\frac{(\epsilon + 2)d\epsilon/dT - (\epsilon - 1)d\epsilon/dT}{(\epsilon + 2)^2} = A \frac{dN}{dT} \tag{12-234}$$

This simplifies to

$$\frac{3d\epsilon/dT}{(\epsilon + 2)^2} = A \frac{dN}{dT} \tag{12-235}$$

The factor A is reexpressed, using Equation 12-233, as

$$A = \frac{\epsilon - 1}{N(\epsilon + 2)} \tag{12-236}$$

This is substituted into Equation 12-235:

$$\frac{3d\epsilon/dT}{(\epsilon + 2)^2} = \frac{\epsilon - 1}{N(\epsilon + 2)} \frac{dN}{dT}$$

And, since $(\varepsilon + 2)$ appears on both sides of this equation,

$$\frac{3d\epsilon/dT}{\epsilon + 2} = \frac{\epsilon - 1}{N}\frac{dN}{dT} \tag{12-237}$$

For the case of perovskites, $\varepsilon \gg 1$ so that $(\varepsilon+2) \simeq (\varepsilon-1) \simeq \varepsilon$. This approximation gives

$$\frac{3d\epsilon/dT}{\epsilon} \simeq \frac{\epsilon}{N}\frac{dN}{dT} \tag{12-238}$$

It will be recognized that dN/NdT is the temperature coefficient of volume expansion, α_v, and that $\alpha_v \simeq 3\alpha_L$, where α_L is the temperature coefficient of linear expansion (see Section 4.3, Volume I). When contraction occurs during cooling in the paralectric state, i.e., above the Curie temperature, T_c, Equation 12-238 may be written as

$$\frac{3d\epsilon/dT}{\epsilon^2} \simeq -\alpha_V \simeq -3\alpha_L$$

Or, upon rearrangement, and noting that the factor 3 vanishes,

$$\frac{d\epsilon}{\epsilon^2} \simeq -\alpha_L dT \tag{12-239}$$

The integration of Equation 12-239 results in

$$\epsilon^{-1} \simeq \alpha_L(T - T_c) \tag{12-240}$$

where the Curie temperature enters as the constant of integration, assuming that α_L is a constant. Equation 12-240 may be reexpressed in the Curie-Weiss form by inverting it to obtain, for perovskites in the paraelectric state,

$$\epsilon \simeq \frac{1/\alpha_L}{T - T_c} \tag{12-241}$$

The relatively small values of α_L result in large values of $1/\alpha_L$, the Curie constant. This also accounts for the large values of ε of most ferroelectric materials. However, the assumption of constancy of α_L makes Equation 12-241 most accurate for $T \gg T_c$, since α_L is a function of temperature (Equations 4-132 and 4-135).

Small changes in the value of $4\pi N\alpha/3$ in Equation 12-23 also can have significant effects upon the value of ε in Equation 12-241. The elimination of A by means of Equation 12-236 minimizes this difficulty in Equation 12-241. The Lorentz field (Equation 12-25) also drops out of the equation. This factor only is approximate since it does not account for the portion of the reaction field parallel to the external field (Equation 12-49). The omission of this factor can lead to a situation in which highly mobile polar molecules, such as in liquids, could show ferroelectric behavior; this does not occur. The use of the Onsager cavity field (Equation 12-55) would eliminate this problem.

The response of ferroelectric crystals at low frequencies (long wavelengths) in the infrared range provides interesting insights into their physical behavior. This may be shown starting with Equation 12-83 and using Equation 12-74 for P_i:

$$m\omega_T^2 = K\left[1 - \frac{P_i}{1 - P_e}\right] = K\left[1 - \frac{(4\pi/3)Ne^2/K}{1 - P_e}\right] \tag{12-242}$$

for the condition where ω_T^2 is relatively small. Here, P_e is the sum of the electronic polarizations of both the positive and negative ions so that Equation 12-75 is used in Equation 12-242 in the form

$$m\omega_T^2 = K\left[1 - \frac{(4\pi/3)Ne^2/K)}{1-(4\pi/3)N\alpha}\right] \qquad (12\text{-}243)$$

where $\alpha = \alpha(+) + \alpha(-)$. It will be recalled that K is the short-range interionic force constant (Equation 12-69). The second term in the brackets of Equation 12-243 is influenced by E_{loc} in the absence of an external field. It results from the low-frequency phonons in the transverse mode as they travel through the lattice. These have the effect of increasing the ionic displacements described in Section 12.5.3. The result is that the amplitudes are increased, the frequencies are lowered and ω_T^2 is decreased.

For the case where ω_T^2 becomes relatively small, $m\omega_T^2$ may be approximated as being negligible so that Equation 12-243 becomes

$$K \simeq \frac{(4\pi/3)Ne^2}{1-(4\pi/3)N\,\alpha} \qquad (12\text{-}244)$$

This provides a means for the approximation of K. However, should $\omega_T \rightarrow 0$, the force constant and the electronic interactions would be equal because α represents the combined electronic polarizabilities of both positive and negative ions. This condition results in very small values of K so that the short-range force is ineffective, the long-range coulombic forces predominate, and the lattice is relatively weakly bound. This suggests that a phase change could take place to form a more stable configuration. However, when a lattice transition (or phase change) occurs, which gives rise to paraelectric properties changing to ferroelectric properties after the phase change, $\omega_T^2 \neq 0$ and $\varepsilon \neq \infty$ (Equation 12-91) during the transition.

The destruction of the spontaneous polarization in heating through the Curie temperature may be a first-order transition and heat is absorbed in the randomization of the dipoles. This energy increment appears as a "spike" on the heat capacity curves at T_c. The behavior is analogous to that discussed in Section 9.4 and shown in Figure 9-10 (Volume II).

The resulting increase in entropy during many transitions is approximately 1 cal/mol/deg, and is not a function of T_c. This indicates that the lattice changes might take place by ordering and disordering. In contrast to this, the entropy increment which is observed in the transition from one ferroelectric phase to another, such as is the case for $BaTiO_3$, is of the order of 0.1 cal/mol/deg, or less. This difference would indicate that the vibrational entropies are significant in these small lattice changes from a ferroelectric crystal structure to a paraelectric lattice type.

The spontaneous polarization may approach zero is a continuous manner, as in Figure 12-18a, when the dipole alignment decreases since $P/Np_{d(max)} \rightarrow 0$ as $T \rightarrow T_c$ (Equation 12-227b). Where the domains are easily reversed, the spontaneous polarization may approach zero discontinuously. This also can occur upon the completion of a lattice transformation (Figure 12-18b). Equations of the same type as Equation 12-241 give good approximations for ε for the paraelectric state. Here $1/\alpha_L$ usually is given as a constant.

12.12.2. Structures and Properties of Some Ferroelectric Crystals

$BaTiO_3$ provides a good example of the changes which are to be expected in ferroelectric properties which result from lattice transitions. It has the simplest structure and more is known about it than most ferroelectrics. Above 120°C it has a cubic lattice and a symmetrical array of ionic charges; it is paraelectric in this state. The cubic lattice transforms to a tetragonal lattice with $c/a \simeq 1.01$ at 120°C. This small distortion results in a slight displacement of the central Ti ions with respect to the surrounding

O ions (Figure 12-14). The resulting ferroelectric polarization is parallel to the c axis and has a value of $P_s \simeq 18 \times 10^{-2}$ coul/m^2. This tetragonal lattice exists down to about 0°C, at which temperature it transforms to an orthorhombic lattice. Here a discontinuous decrease occurs in the polarization. This lattice may be thought of as a distorted cubic lattice which has been slightly expanded along one face diagonal and compressed slightly along the other face diagonal. These distortions are parallel to the orthorhombic axes. This also results in a displacement of the Ti ion with respect to the O ions. Such distortions cause ionic displacements which are parallel to each other in the c direction. It is assumed, but not as yet proven, that the polarization is parallel to the expanded axis of the orthorhombic structure cell. This assumption is based upon the polarization of the tetragonal lattice. Further cooling down to about −90°C results in another lattice transformation and a discrete decrease in polarization. There is some question as to whether this lattice is trigonal or rhombehedral. In either event, the polarization is parallel to the body diagonal.

Each of the lattice transformations which occur at the two lower temperatures show hysteresis effects in changes of both lattice types and polarizations; transformations occurring upon cooling occur at higher temperatures than those which take place upon heating.

Other compounds of this class may or may not undergo crystalline transformations. $LiTaO_3$, which does not change its crystal structure, has values of $P_s \simeq 20 \times 10^{-2}$ coul/m^2 and $T_c = 450$°C. $KNbO_3$ can be made to transform its crystal structure. When it is in the same tetragonal form as $BaTiO_3$ it has values of $P_s \simeq 30 \times 10^{-2}$ coul/m^2 and $T_c \simeq 410$°C.

Different types of ionic displacements may take place in lattices which do not result in spontaneous polarizations. Such materials have crystalline configurations such that adjacent arrays of dipoles either are aligned in opposite directions or that the displacements of positive ions in adjacent arrays oppose and cancel each other. Materials of these types are antiferroelectrics.

All of the ferroelectrics and antiferroelectrics described thus far are ionic compounds; their bonding either is completely or almost completely ionic. However, the nondirectional character of ionic bonds (Section 10.6.1) cannot explain the highly directional ionic displacements required to explain ferroelectric behavior, especially those which occur as a result of lattice transformations. Some degree of covalency of the bonding component has been considered to be present and to have an important influence upon the octahedral oxygen ions (see Section 10.6.3 and Figure 10-12). This provides a basis for the lattice transformations and permits the corresponding changes in polarization to be considered to result from changes in the degree of covalent bonding present.

Where a positive ion occupies a central position in a cubic lattice in which oxygen ions are symmetrically arrayed about it, no dipole moment can exist. It is probable that little or no covalent bonding exists in this case. However, such compounds need not be completely ionic. Transformations to the ferroelectric state can be explained in terms of the relative amounts of ionic and covalent bonding (see Sections 10.6.7 and 10.6.8).

The KDP (potassium dihydrogen phosphate) type ferroelectrics crystallize in complex-diamond-type, tetragonal unit cells. The principle structure elements are tetrahedra formed by the $(PO_4)^{-4}$ ions. In the case of KH_2PO_4, the PO_4 tetrahedra are centrally located upon the positions of a diamond-type tetragonal lattice similar to Figures 10-17a and b. However, instead of the C-C bonds, each PO_4 ion is bonded to four nearest PO_4 neighbors by a hydrogen bond. These hydrogen bonds are made between an O ion in the upper part of one tetrahedron and another O ion in the lower part of an adjacent tetrahedron, with a bond length of about 2.5 Å. In the paraelectric state the H ions are centrally located between the PO_4 ions and form bonds analogous to C-C bonds in the diamond-cubic structure cell.

The positions of K ions in the diamond-tetragonal lattice formed by the PO_4 tetrahedra are as tabulated:

Base-centered positions: 1/2, 1/2, 0; 1/2, 1/2, 1
Cell-edge positions: 0, 0, 1/2; 1, 0, 1/2; 1, 1, 1/2; 0, 1, 1/2
Cell-face positions: 1, 1/2, 1/4; 0, 1/2, 1/4; 1/2, 1/2, 3/4; 1/2, 0, 3/4

In the ferroelectric state the H ions (protons) take preferred positions within the lattice of a domain. Two H ions will be very close to each PO_4 ion. In one domain they may be near upper O ions; in another, near lower ions. The H ions switch from one of these positions to the other when polarization reverses; the difference between the two positions is about 0.4 Å.

The major contribution to the polarization arises from the close association of the H and PO_4 ions. This causes the polarization of the PO_4 ions parallel to the c axis. This results in ε being between 10^4 and 10^5 in the c direction as compared to about 10^2 in the a direction at $T \simeq -150°C$. An additional component of polarization also results from this close associatipn; the P and K ions appear to be forced away from those O ions in close proximity with an H ion. The importance of the role of the interaction of H and PO_4 ions is illustrated when deuterium is substituted for the hydrogen. This has the effect of nearly doubling T_c for KDP (approximately from −150°C to about −60°C) while P_s decreases (from about 6×10^{-2} to 5×10^{-2} coul/m²).

Most KDP compounds have P_s in the range from about 5×10^{-2} to 6×10^{-2} coul/m². It should be noted that engineering applications of many KDP-type ferroelectric materials are limited by their low Curie temperatures. Most of these are less than 150 K. Many of these substances have values of ε between 10 and 10^2 near room temperature.

Rochelle salt is a hydrated tartrate with the chemical formula $NaKC_4H_4O_6 \cdot 4H_2O$. It has a complex orthorhombic crystal structure when it is polarized. The ferroelectric axis parallel to the a edge is very high, and unlike KDP compounds, its dielectric constant is a function of temperature. Two sharp T_c peaks occur at about −18°C and +23°C at which ε is of the order of 10^3. The dielectric constants for b and c directions show a small, linear response to temperature. These have values of ε of about 10 in the neighborhood of room temperature. In the range of −18°C to +23°C its structure is monoclinic. The structure is orthorhombic above and below this range. At temperatures below −120°C the dielectric constant parallel to the a direction becomes linear and approaches the values of those of the b and c directions.

The crystal structure of Rochelle salt is extremely complicated. Several explanations have been proposed to describe the small displacements responsible for its ferroelectric behavior. It has been thought that hydrogen bonds and water of hydration may be important factors because the substitution of D for H causes small but significant shifts in the two Curie temperatures.

Other tartrates such as $LiNH_4$-tartrate and LiTl-tartrate each contain only one water molecule. Their structures are similar to that of Rochelle salt. $LiNH_4$-tartrate shows only very small changes in T_c when D is substituted for H.

Capacitors employing ferroelectrics, usually perovskites, can be made to change their values of ε depending upon changes in the magnitude and direction of the electric field. Under these conditions they can be made to behave like varactors (Section 11.9.6.1). Such capacitors also may be used as radiofrequency power amplifiers. In this application the electric field changes ε and this, in turn, changes the power supply. Changes in the field also modulate the frequency.

Very thin ferroelectric crystals can be used for computer memories. Parallel sets of conductors are deposited upon each side of the crystal. The set of conductors on one side is perpendicular to the set on the other side. Voltages across a given pair of conductors, one on either side of the ferroelectric, will polarize a given small volume in the ferroelectric crystal. The polarization in this small volume will be in one of two opposite directions, depending upon the field between the conductors. The response of the polarized volume to a potential difference caused by a pulse in the conductors above and below it depends upon its initial polarization direction and the electric field induced by the pulse. In one direction of polarization, the response will be a pulse of current, in the other direction it will be a pulse of voltage. The memory packing density of ferroelectrics is comparable to that of ferrites, but their rate of switching is slow comparred to ferrites.

12.13. PIEZOELECTRICITY

The induction of a dipole moment in a dielectric by an applied field causes a displacement of the ions and the dimensions of the crystal are changed. This reaction of the solid to the applied field is known as electrostriction. The effect of this polarization is to strain the crystal. This may be approximated for a single crystalline direction in these highly anisotropic crystals by

$$\epsilon(P) = \sigma(P)s + \bar{E}d \qquad (12\text{-}245)$$

in which $\varepsilon(P)$ is the elastic strain, $\sigma(P)$ is the mechanical stress, s is an elastic compliance constant for a given electric field, and d is a piezoelectric strain coefficient. Both s and d are one-dimensional parameters in Equation 12-245. In the most general case the 36 coefficients of d and E form a third-rank tensor for P and s also may have as many as 36 values in the six generalized equations of Hooke's law which include the effects of the electric field.

The anisotropic nature of s and d show that the symmetry of crystals is an important consideration. A crystal has a center of symmetry if two equal and opposite vectors drawn from a central point in the structure cell ends at ions of the same species. The ions of crystals with symmetrical structure cells move in a uniform way and preserve their symmetry under the influence of a mechanical force, either tension or compression. Thus, no changes in polarization take place in these crystals as a result of applied stress. Where the structure cell of a crystal has some degree of asymmetry, the dipole moment may be decreased as a result of mechanical compression or increased by tension. Of the natural 32 crystal types, 21 are unsymmetrical to some degree. However, many materials of these types have piezoelectric properties which are too small to be of significance.

It is interesting to note that materials with 10 of these lattice types are pyroelectric. These pyroelectric materials change their polarizations with changes in temperature. This results from changes in their dipole moments which occur upon the thermal expansion or contraction of the crystal. All pyroelectric materials show piezoelectric properties.

Ferroelectric crystals, by their very nature, are highly unsymmetrical, belong to the class of pyroelectric materials, and show large piezoelectric responses. The application of external mechanical forces to such crystals changes the distances between the ions and alters their dipole moments. The relatively high degree of anisotropy of these crystals results in the complex relation between strain and polarization for the various crystalline orientations because of the true nature of s and d in Equation 12-245. Changes in the dipole moments induced by mechanical means result in changes in their polarization. The amount of change in a given direction depends upon the degree of anisotropy of the crystal. Any materials which change polarizations in this way are piezoelectric.

The polarization is approximated for a given crystalline direction by

$$P \simeq \sigma(P)d + \overline{E}\chi_e;\ \chi_e = P/\overline{E} \qquad (12\text{-}246)$$

Piezoelectric materials can act as transducers because a mechanical stress induces an electric field and, conversely, an electric field induces a mechanical stress, or dimensional change. Several mechanical modes of vibration may occur naturally in piezoelectric crystals because s and d vary with the crystalline direction. The geometry of the structure cell, the elastic constants, the dimensions of the crystal and the orientation of the crystal with respect to the applied field will determine the frequencies of vibration. The application of an alternating field across a piezoelectric crystal will cause it to oscillate. The maximum amplitude of vibration is obtained when the alternating field is at the resonant frequency of a given crystal. This is determined by the factors noted previously, especially s and d which are anisotropic.

Crystals of Rochelle salt (see Section 12.12) have large values of d and are useful as transducers. Other ferroelectric materials such as $BaTiO_3$ are used as transducers in the polycrystalline form rather than as single crystals. Solid solutions of $PbZrO_3$ and $PbTiO_3$, which have the same crystal structure as $BaTiO_3$, are also used as transducers, especially in phonograph pick-ups. These have higher Curie temperatures. This class of ceramics is conditioned by heating above the Curie temperature and cooling under the influence of a magnetic field to give the desired polarization.

Quartz crystals, which are not ferroelectric, are widely used as oscillators in electronic applications. They also are used in the generation and detection of ultrasonic waves. One important application of this in the inspection of solids for the presence of internal flaws. Such crystals are produced as thin slices in which the basal planes are parallel to the faces of the slice. The application of an electric field perpendicular to the faces will cause them to expand or contract parallel to the field. The strong covalent bonding of quartz crystals results in stable values of s and d. This is responsible for the high degree of constancy of these crystals and their wide application as frequency generators and standards.

Piezoelectric materials also are used as delay lines in communications circuitry. An oscillating electric signal is transformed into an acoustic wave at one end of a rod of piezoelectric material. These accoustic waves will vary with the crystal structure, crystalline orientation of the piezoelectric and the frequency of the electric signal. The maximum velocity of the accoustic wave in the piezoelectric is that of the speed of sound in the material. This is very slow compared to the velocity of an electric signal in a metallic conductor ($\sim 10^8$ cm/sec). The slowed signal, in the form of an accoustic wave, is then reconverted into an electric signal at the other end of the rod where it enters a metallic conductor again.

12.14. DIELECTRIC INSULATING MATERIALS

One of the most important engineering applications of dielectric materials is their very extensive use as electrical insulators. A nonpolar material would always be chosen for electrical insulation purposes if dielectric properties were the sole consideration. Such a material would guarantee high insulation resistance and high dielectric strength, as well as low dielectric losses. However, many other considerations must be taken into account in the selection of a dielectric for a given application. These include mechanical properties, variations of the mechanical, physical, and chemical properties as functions of temperature, the stability of the electrical properties in the temperature and frequency ranges of proposed use and the environment of that application.

The large number of classes of dielectric materials used for electrical insulation purposes, as well as the number within a given class, makes it impossible to provide complete descriptions of the properties of all these materials.[13,14] A general description of the properties of some of the most widely used types of electrical insulators is given here. The units used in each case are those generally accepted by those using a given class of material.

Impregnated papers have long been used as insulators. The impregnation replaces the air within the paper with a material which has a higher dielectric strength and which also protects the paper from moisture absorption. At low temperatures the loss factor is largely determined by the paper itself and is a function of the density of the paper alone. The loss factor varies from <0.002 to ~ 0.003 for densities between 0.7 and 1.1 g/cm^3 at room temperature. This property goes through a minimum near 60°C and increases to about 0.0035 at 100°C. The loss factor is virtually unaffected by the field strength up to about 200 kV/cm. Higher loss factors may result from poor impregnation, contamination, electrode defects, or moisture absorption. The dielectric constant is usually greater than 2.5. Up to about 1,000 Hz the dielectric constant decreases slightly, but is nearly constant beyond that frequency. Well-impregnated papers have electrical resistivities of about 10^{17} Ω-cm.

Hydrocarbon insulating oils are used for impregnating paper and for filling hollow-core cables. Their properties are determined by the crude oils from which they originate and the refining processes employed in their production. The amounts of polar and/or conducting phases in the oils are reflected in their electrical properties. The loss factors vary between 0.00001 and 0.00005, with electrical resistivities between 10^{15} and 10^{17} Ω-cm. Their dielectric constants average about 2.3. Dielectric breakdown occurs at about 400 to 500 kV/cm. Dielectric strengths of up to 1,000 kV/cm have been obtained for specially prepared oils.

Chlorinated hydrocarbons, such as chlorinated diphenyls and benzines, are used as insulations for capacitors and transformers. These are more costly than the oils, but have advantages over them. These materials are neither flammable nor explosive under arcing conditions. Neither phosgene nor chlorine are given off. However, hydrogen chloride, which is very corrosive, can be evolved. Thus, these compounds cannot be used in switchgear, while the hydrocarbon oils are used for this purpose. They do, however, have dielectric strengths comparable to the hydrocarbon oils and are stable under normal conditions. They have acceptable loss factors. The properties of this class of materials vary widely with their molecular structures. their dielectric constants are relatively high and range from about 4 to 6. It should be noted that severe restrictions, and in some cases prohibitions, have been placed upon the use of these compounds because they can constitute health and environmental hazards.

Synthetic and natural rubbers are widely used as insulators for wires and cables. These have a high degree of flexibility, wear resistance, and toughness, resist chemical attack, as well as having good dielectric properties. Their polymeric structures greatly influence the properties of these materials. Most of the nonpolar materials have small dipole moments. The vulcanization of rubbers increases their dipole moments, but most of these materials are not highly polar. Electronic, atomic, and some ionic polarizabilities are present. Rubber polymers containing polar arrays show considerable orientation polarization and corresponding losses in alternating fields. Their relaxation times and loss factors vary more than do those with smaller molecules. Pure, unvulcanized rubber has a dielectric constant of ~ 2.4 and a loss factor of ~ 0.003 at room temperature. Vulcanized rubber has a dielectric constant of $\sim$2.6 and a loss factor of $\sim$0.04. Both of these sets of data were taken at 1000 Hz. Breakdown takes place between 300 and 600 V/mil. The electrical resistivities of these materials range from about 10^{14} to 10^{16} Ω-cm.

Plastics, or synthetic high polymers, have good combinations of thermal, mechanical, and chemical properties. The dielectric constant of polythene is about 2.3 at 20°C and is independent of frequency from 50 to 10,000 Hz. Its loss factor is between 0.0001 and 0.0002 up to 1000 Hz. Dielectric breakdown occurs at about 400 kV/cm at 50 Hz and 20°C. This type of material shows very little tendency toward water absorption. Polystyrene and polyvinyl chloride have properties which are fairly representative of this class of materials. The dielectric constant of polyvinyl chloride is about 3.0, and its loss factor is about 0.02 at frequencies up to 1 MHz. Breakdown occurs between 50 and 200 V/cm. Mylar polyester film is very useful for capacitor applications at moderate temperatures, since its mechanical properties vary only slightly from −20 to about 80°C. Its elastic modulus and tensile strength decrease rapidly above this temperature. Its dielectric constant is $\sim$3.1 and its loss factor is $\sim$0.005 at 25°C. Breakdown occurs between 8000 and 12,000 V/cm.

The phenolic resins are used with fillers such as wood flour and cotton flock to make them less brittle. Powdered mica and asbestos are included among many other filler materials used. Their dielectric properties vary depending upon the filler material. Their dielectric constants range from 5 to 10 at 60 Hz, their loss factors lie between 0.03 and 0.30, and their electrical resistivities vary from 10^{9} to 10^{14} Ω-cm.

Many resins are used as coatings for conductors or as impregnants for such components as coil or motor windings. Where solvents are used in the application of these materials, it is virtually impossible to produce completely coated windings. Bubbles, which are produced by the evaporating solvents during the curing process, leave portions of the conducting surfaces uncoated. This problem may be avoided by using resins which do not require the use of solvents. Coatings of this type act as barriers to moisture or to other undesirable environmental factors. In addition, as impregnants, they strengthen components such as windings by cementing the conductors together and by effectively encapsulating them.

Resin-bonded laminates are made with paper, cloth, asbestos, and other materials. The properties of laminates depend upon their components. These have dielectric constants varying between 4.5 and 7.0, and loss factors of from 0.03 to 0.10 at 1 MHz. Paper laminates have dielectric strengths of about 500 V/mil. Fabric laminates undergo breakdown from about 150 to 400 V/mil. Most laminates have electrical resistivities of the order of 10^{15} to 10^{16} Ω-cm.

Silicones show good dielectric properties over a large range of frequencies. Their physical properties are relatively constant compared to organic materials and they are chemically inert. Silicone rubbers are used primarily for electrical insulation purposes. Their dielectric constants are only slightly affected by temperature and range from about 3 to 10 when measured at 10 kHz. Their loss factors range up to about 0.03 at this frequency. Breakdown occurs between 1 and 10 kV/cm.

Ceramics have their largest number of applications as insulators. They also are used in some capacitors. Their dielectric constants and loss factors increase relatively rapidly with both temperature and frequency. The higher the frequency the smaller is the effect of temperature upon their dielectric constants and loss factors. Typical ranges of the properties of these materials are: dielectric constant, 4.5 to 9; loss factor, 0.003 to 0.02; and dielectric strength, 3 to 50 kV/mm. Ceramics, including porcelains, are most frequently used for high-voltage applications. They have good mechanical strength and dimensional stability. They are chemically stable and are widely used out of doors because of their good corrosion resistance and their low permeability to gases. They also have low losses. Most ceramic materials are relatively inexpensive.

Glasses are mixtures of SiO_2 and other oxides. They are very viscous and noncrystalline in the molten state. Glasses retain their liquid-like, structural noncrystallinity, and, at room temperature, are really supercooled liquids. Their constitution is such that little limitation is placed upon the number and quantity of the oxides which may be added to the silica base. As would be expected, the properties of glasses vary considerably with composition.

The most common glasses are the soda-lime-silica (chiefly SiO_2, CaO, and Na_2O) glasses. Other types include borosilicate and lead glasses. Glasses have electrical resistivities which range from 10^{10} to 10^{20} Ω-cm. The little electrical conduction which does take place is ionic. This results mainly from the diffusion of sodium ions in the highly distorted silica "lattice". Other cations, such as potassium, calcium, and magnesium, are larger in size and have smaller mobilities than the sodium ions; therefore, they contribute only in a minor way to the electrical conductivity. These factors explain the high electrical resistivities and negative temperature coefficients of resistivity of glasses. A typical glass can show a decrease in resistivity of from about 10^{16} Ω-cm at room temperature to to 10^{9} Ω-cm at 300°C. This behavior is manifested because of the effects of the decrease in the rate of transfer of the thermal energy.

The dielectric constants of commercial glasses lie between 3.7 and 10. The common soda-lime-silica glasses have dielectric constants of about 7 to 8. Their dielectric properties result primarily from electronic polarizability with minor contributions from atomic displacements which are induced by the electric field. Thus, the dielectric constants of these glasses are virtually independent of the frequency.

The dielectric losses in glasses are a function of frequency. The loss factor is about 0.02 at 10^{3} Hz, goes through a minimum of 0.01 at about 10^{7} Hz, and then sharply increases to about 0.02 at 10^{10} Hz. It has been considered that the increased losses at the higher frequencies is a result of vibrational losses. The dielectric strengths of these glasses range from about 3000 to 5000 kV/cm under laboratory conditions. A value of 1700 kV/cm often is used for engineering applications. This figure decreases rapidly with temperature and is about half of the room-temperature value at 100°C. This results from the large, negative temperature coefficient of resistivity and other thermal effects noted earlier.

Mica is a generic name for a group of natural minerals with laminar structures. This permits them to be divided into thin sheets. These minerals have low dielectric losses, high dielectric strengths, resistance to moderate temperatures, and good mechanical strengths when in the form of thin sheets. It should be noted that the mica sheets tend to split into many thin laminae when heated to too high a temperature. The sheets have poor flexibility and cannot be moulded. Micas are available in various grades depending upon their compositions and impurity contents. The best grades of mica have low impurity contents and are transparent and smooth. The dielectric strength of high-quality mica (muscovite) is greater than 1500 V/mil. Its dielectric constant is about 5.4 and it is nearly independent of frequency from 1 kHz to 3000 MHz. The loss factor is nearly constant over this range of frequencies; it varies between 0.0001 and 0.0002.

Both natural and synthetic fibers and textiles are used for insulation purposes. Many textiles are durable, strong, and flexible. Their properties vary widely, depending upon their composition, and are strongly affected by their environment. When these are perfectly dry, their dielectric constants are about 2 over a frequency range of 10^{2} to 10^{7} Hz. When these materials are subjected to high-humidity conditions, their dielectric constants vary with frequency; they initially range from about 5 to 7 and drop steeply

Table 12-5
DIELECTRIC CONSTANTS OF SOME PLASTICS AND RUBBERS

Material	Temp (°C)	Frequency (Hz) 1 × 10³	1 × 10⁶	1 × 10⁸
Phenol-formaldehyde	25—27	5.15—8.61	4.45—5.05	4.1—4.5
	57	6.35	4.90	4.5
	88	8.5	5.2	4.7
Melamine-formaldehyde	24—28	6.0—6.90	5.82—6.20	5.5—5.55
	57	6.95	5.40	4.90
	88	11.8	6.0	5.5
Nylon 66	25	3.75	3.33	3.16
Cellulose Acetate	26	3.50—4.48	3.28—3.90	3.05—3.40
Silicone resins	25	3.79—3.91	3.79—3.82	3.82
Polyethylene	23	2.26	2.26	2.26
Vinylite VU	24	5.65	3.30	2.80
Polyvinyl chloride	25	4.55(1×10⁴)	3.3	—
Lucite	23	2.84	2.63	2.50
Plexiglas	27	3.12	2.76	—
Polystyrene	25	2.54—2.56	2.54—2.56	—
Hevea				
Vulcanized	27	2.94	2.74	2.46
Compound	27	36	9	6.8
Buna S	20	2.66	2.56	2.52
Neoprene	24	6.60	6.26	4.5
Silicone rubber	25	3.12—3.30	3.10—3.20	3.06—3.18

Abstracted from Weast, R., Ed., *Handbook of Chemistry and Physics,* 56th ed., CRC Press, Boca Raton, Fla., 1975, E60.

Table 12-6
DIELECTRIC CONSTANTS OF SOME CERAMICS

Material	ε (1 × 10⁶ Hz)	Dielectric strength (V/mil)
Alumina	4.5—8.4	40—160
Corderite	4.5—5.4	40—250
Fosterite	6.2	240
Porcelain		
(Dry process)	6.0—8.0	40—240
(Wet process)	6.0—7.0	90—400
Zircon	7.1—10.5	250—400
Steatite	5.5—7.5	200—400
Titanates (Ba, Sr, Ca, Mg, and Pb)	15—12,000	50—300
Titanium dioxide	14—110	100—210

Abstracted from Weast, R., Ed., *Handbook of Chemistry and Physics,* 56th ed., CRC Press, Boca Raton, Fla., 1975, E60.

and continuously to about 2 over the same range of frequencies. This indicates that the relaxation times can only be explained by a range of relaxation times instead of a single relaxation time. Their loss factors show similar behavior. These lie between 0.003 and 0.02 for dry textiles over the same frequency range. The loss factors generally decrease with frequency under high-humidity conditions. At a given frequency, both the dielectric constants and the loss factors increase as a function of water content.

The electrical conduction mechanism in water-absorbing fibers appears to be a result of ions in solution in the moisture. Surface conduction is thought to occur in those fibers which do not absorb water. The electrical resistivities of dry fibers lie in the range from 10^{11} to 10^{14} Ω-cm. Typical average breakdown strengths of these dry materials lie between 300 and 700 V/mil.

Additional data on some commonly used dielectric materials are given in Tables 12-5 and 12-6.

12.15. PROBLEMS

1. Explain the reasons for the relative orientation polarizability of gases, liquids, and solids in general.
2. Discuss the utility of highly polarizable materials in the design of capacitors. What uses may be made of various dielectric materials which span a large range of polarizability?
3. Explain the decrease in polarization as a function of increasing temperature.
4. Why are the dielectric properties of nonpolar materials less affected by increasing temperature than polar materials?
5. Explain the polarity of the HCl, H_2S, and CCl_4 molecules.
6. Approximate α_e by means of Equations 12-15a and 12-15b and compare the results with the data.
7. Given the following ionic radii (from Laves, F., in *Theory of Alloy Phases,* A.S.M., 1956, 131.): $F^- = 1.36$ Å, $Cl^- = 1.81$ Å, $Br^- = 1.95$ Å and $I^- = 2.16$ Å, estimate α_e using Equations 12-15b. Calculate the average percentage error based upon data in Table 12-2.
8. What percentage of the valence electrons in a water molecule are involved in its polarization if the H-O-H bonds form an angle of 104.5° and their length is 0.958 Å? What information does this reveal about the bonding mechanism?
9. Derive an expression for the Debye length, L_D, cited in Section 11.2, starting with Equation 12-14. Consider the stored energy per ion as being $\varepsilon\bar{E}/2$.
10. Why is it possible to use the Langevin theory for both paramagnetism (Section 8.3.1, Volume II) and for orientation polarization?
11. Show that the Debye equation (Equation 12-28b) gives erroneous results for pure water. Compare the calculated values of α_a and p_d with those given in the tables. Explain the discrepancies. Hint: solve Equation 12-28b simultaneously using data in the neighborhood of room temperature for each equation. Also note that α_e should be used instead of α_a in this case.
12. CCl_4 has a density of 1.63 g/mℓ at room temperature. Calculate its dielectric constant and index of refraction. How does this compare with the experimental value of $n = 1.46$? What is its molar polarization?
13. Diamond crystals have a lattice parameter of 3.57 Å and $n = 2.45$. Calculate the polarizability of C in the diamond-cubic lattice. Compare the result with the datum in Table 12-2. Estimate its "radius".
14. Use the data obtained in problem 12 to obtain the electric susceptibility of CCl_4.
15. Calculate the dielectric constant of a mixture of two nonpolar liquids A and B if their volume fractions are in the ratio 3A:1B and the dielectric constants of A and B are 2.24 and 2.02, respectively.
16. Calculate the Onsager dipole moment of an NaCl molecule at 27°C from its lattice parameter of 5.63 Å and information given in Table 12-4.
17. Use data from problem 16 to calculate the polarizability of NaCl under static conditions.

18. Calculate the percentages of ionic and electronic polarization in NaCl crystals.
19. Calculate the average ionic displacement in the NaCl lattice using data from problem 16.
20. Do the data obtained in problem 17 agree with Equation 12-89? Verify the result with data from Table 12-4. How well do the data used in problem 16 agree with these?
21. Approximate the average vibrational frequency of the ions in an NaCl crystal, under static conditions, using polarizability data from Table 12-2. Check the results using $\omega_T \simeq 3 \times 10^{13}$/sec.
22. Derive an equation for the damping coefficient from Equation 12-116a. Draw a graph of dB/dω (in arbitrary units) as a function of ω/ω_o. Within what range of values of ω/ω_o can the change in B be considered to be essentially constant? Explain the meaning of the behavior of dB/dω as $\omega \rightarrow \omega_o$.
23. (a) Show that Equation 12-139b has a maximum at $\omega\tau = 1$. (b) What is the relationship between ε and ε_p under this condition? (c) What relationship exists between ε and ε_p when Equation 12-139a is minimized? (d) What is the value of Equation 12-139a under this condition® Verify the results by use of Figure 12-8.
24. Why may classical statistics be used to describe conduction mechanisms in dielectric materials?
25. On what basis may it be reasoned that the energy required for the creation of a Schottky pair must be smaller than that for the creation of an interstitial?
26. Why should intrinsic conduction in Laves-type crystals be primarily a result of vacancy migration rather than by interstitials?
27. Why may ion and vacancy migrations transverse to an applied electric field be neglected in analyses of intrinsic and extrinsic conductivities of dielectric crystals?
28. What multiplying factor is required for the application of Equation 12-168 to a CsCl-type lattice?
29. Why is the emphasis placed upon cation vacancies in Equations 12-175, 12-176, and 12-189?
30. Express the total conductivity of a dielectric solid using Equations 12-175 and 12-190 in terms only of cation vacancies.
31. Given that the diffusion coefficient, ionic conductivity, and activation energy for Na^+ in NaCl are approximately 5×10^{-10}cm^2/sec, $9 \times 10^{-5}(\Omega\text{-cm})^{-1}$ and 1.94 eV, respectively, calculate (a) the mobility and (b) the number of Na^+ engaging in this process. Use the Boltzmann statistics and information from problem 16 to check (b).
32. Use information from problems 16, 21, and 31 to approximate the conductivity due to the cation vacancies in an NaCl lattice. What percentage of the ionic conductivity does this represent? Hint: use $\Delta F = F_p$.
33. If an addition of 3×10^{11} divalent cations is made to a pure NaCl crystal whose dimensions are 1 cm × 1 cm × 1 cm, approximate the conductivity of the crystal at 900°C. Use μ from problem 31. Why is this permissible?
34. Derive a relationship for the density of the space charge in a dielectric material which includes the effects of back diffusion.
35. A dielectric material is at the threshold of breakdown in a field of 10^7 V/cm. The approximate average carrier mobility is 10^3cm^2/statvolt sec. Calculate the minimum rate of removal of thermal energy required to avoid breakdown. Explain the order of magnitude of the mobility.
36. Use graphical means to obtain a curve for the relative ferroelectric polarization as a function of temperature.
37. Use the Onsager cavity field (Equation 12-55) to obtain a correction factor for

the coefficient β in the Lorentz internal field to provide more accurate results from Equations 12-225 and 12-227. Hint: $\varepsilon \gg 1$.

38. $BaTiO_3$ has an approximate lattice parameter of 4 Å (consider it as being cubic) and $T_c = 120°C$. Its dielectric constant is $\simeq$7000 at 125°C. Calculate the approximate values of β and p_d per structure cell. Is the correction factor found in problem 37 applicable here?
39. Given that the dielectric constant of $BaTiO_3$ is $\simeq$645 at 320°C, estimate the extent to which the polarization catastrophe has taken place.
40. Using the data in problem 39 and $\alpha_L \simeq 0.8 \times 10^{-5}/°C$ at 320°C, approximate T_c. Explain the difference between this value and the experimental value given in problem 39.
41. Calculate the short-range interionic force constant of a ferroelectric at room temperature if its reduced mass, lattice constant, α_e, and ω_T are 55.7×10^{-23}g, 4 Å, 9.3×10^{-24}cm^3, and 2.1×10^{13}/sec respectively. Estimate the value of ω_T at T_c which would cause $K \rightarrow 0$.
42. Explain how variations in the degree of covalent bonding in a lattice in which the bonding is primarily ionic could account for the differences in ferroelectric behavior as a result of allotropic changes.
43. Give the reasons for the use of high-purity mica in precision capacitors.
44. Explain the dielectric behavior of a moisture-containing textile as a function of frequency.

12.16. REFERENCES

1. **Dekker, A. J.,** *Solid State Physics,* Prentice-Hall, Englewood Cliffs, N. J., 1957.
2. **Kittel, C.,** *Introduction to Solid State Physics,* 3rd ed., John Wiley & Sons, New York, 1966.
3. **Böttcher, C. J. F., et al.,** *Theory of Electric Polarization,* Vol. 1, 2nd ed., Elsevier, Amsterdam, 1973.
4. **Zheludev, I. S.,** *Physics of Crystalline Dielectrics,* Vols. 1 and 2, Plenum Press, New York, 1971.
5. **Daniel, V. V.,** *Dielectric Relaxation,* Academic Press, New York, 1967.
6. **O'Dwyer, J. J.,** *The Theory of Electrical Conduction and Breakdown in Solid Dielectrics,* Clarendon Press, Oxford, 1973.
7. **Hutchison, T. S. and Baird, D. C.,** *The Physics of Engineering Solids,* John Wiley & Sons, New York, 1968.
8. **Mott, N. F. and Gurney, R. W.,** *Electronic Processes in Crystals,* Dover, New York, 1964.
9. **Hill, N. E., et al.,** *Dielectric Properties and Molecular Behavior,* Van Nostrand Reinhold, 1969.
10. **Kanzig, W.,** Ferroelectrics and Antiferroelectrics, in *Solid State Physics,* Vol. 4, Seitz, F., and Turnbull, D., Eds., Academic Press, New York, 1957.
11. **Burfoot, J. C.,** *Ferroelectrics, An Introduction to the Physical Principles,* Van Nostrand, New York, 1967.
12. **Fatuzzo, E., and Merz, W. J.,** Ferroelectricity, in *Selected Topics in Solid State Physics,* Vol. VII, Wohlfarth, E. P., Ed., John Wiley & Sons, New York, 1967.
13. **Birks, J. B.,** *Modern Dielectric Materials,* Academic Press, New York, 1960.
14. **Saums, H. L. and Pendleton, W. W.,** *Materials for Electrical Insulating and Dielectric Functions,* Hayden, Rochelle Park, N.J., 1973.
15. **Slater, J. C. and Frank, N. H.,** *Introduction to Theoretical Physics,* McGraw-Hill, New York, 1933, 441.
16. **Jonker, G. H. and van Santen, J. H.,** *Science,* 109, 632, 1949.

APPENDIX A

USEFUL PHYSICAL CONSTANTS

Constant	Symbol	Value CGS	Value SI
Electron rest mass	m	9.11×10^{-28} g	9.11×10^{-31} kg
Electron charge	e	4.80×10^{-10} esu	1.60×10^{-19} coul
Planck's constant	h	6.63×10^{-27} erg sec	6.63×10^{-34} J sec
Planck's constant/2π	$\hbar$	1.05×10^{-27} erg sec	1.05×10^{-34} J sec
Boltzmann's constant	k_B	1.38×10^{-16} erg/K	1.38×10^{-23} J/K
	k_B	8.63×10^{-5} eV/K	
Electron volt	eV	1.60×10^{-12} erg	1.60×10^{-19} J
Electron volt/molecule	eV/a	23.06 kcal/mol	
Gas constant	R	1.987 cal/(mol K)	8.31 J/(mol K)
Avogadro's number	N, N_A	6.02×10^{23}/mol	6.02×10^{26}/kmol
Atomic mass unit	amu	1.66×10^{-24} g	1.66×10^{-27} kg

APPENDIX B

CONVERSION OF UNITS

Unit	Symbol	Conversion operation	Resulting units
Electrical potential (volt)	V	V/300	statvolt
Electrical current (amp)	A	$A/(3 \times 10^9)$	statamp
Electric field (V/cm)	$\bar{E}$	$\bar{E}/300$	statvolt/cm
Conductivity [(Ω-cm)$^{-1}$]	σ	$\sigma \times 9 \times 10^{11}$	esu conductivity
Resistivity (Ω-cm)	ϱ	$\varrho/(9 \times 10^{11})$	esu resistivity
Mobility [cm^2/(V sec)]	μ	$\mu \times 300$	cm^2/statvolt sec
Current density (amp/cm^2)	j	$j \times 3 \times 10^9$	statamp/cm^2
Magnetic flux density (Weber/m^2)	T	$T \times 10^4$	gauss
Magnetic field strength (amp turns/m)	H	$H \times 4\pi \times 10^{-3}$	oersted
Thermal conductivity [cal/(cm sec°C]	$\varkappa$	$\varkappa \times 422$	Watt/(m°K)

INDEX

A

B

C

D

E

F

G

H

I

J

K

L

M

N

O

P

Q

R

S

T

U

V

W

X

Y

Z